210
Advances in Polymer Science

AF798258

Advances in Polymer Science
Recently Published and Forthcoming Volumes

Wax Crystal Control · Nanocomposites Stimuli-Responsive Polymers
Vol. 210, 2008

Functional Materials and Biomaterials
Vol. 209, 2007

Phase-Separated Interpenetrating Polymer Networks
Vol. 208, 2007

Hydrogen Bonded Polymers
Volume Editor: Binder, W.
Vol. 207, 2007

Oligomers · Polymer Composites Molecular Imprinting
Vol. 206, 2007

Polysaccharides II
Volume Editor: Klemm, D.
Vol. 205, 2006

Neodymium Based Ziegler Catalysts – Fundamental Chemistry
Volume Editor: Nuyken, O.
Vol. 204, 2006

Polymers for Regenerative Medicine
Volume Editor: Werner, C.
Vol. 203, 2006

Peptide Hybrid Polymers
Volume Editors: Klok, H.-A., Schlaad, H.
Vol. 202, 2006

Supramolecular Polymers Polymeric Betains · Oligomers
Vol. 201, 2006

Ordered Polymeric Nanostructures at Surfaces
Volume Editor: Vancso, G. J., Reiter, G.
Vol. 200, 2006

Emissive Materials · Nanomaterials
Vol. 199, 2006

Surface-Initiated Polymerization II
Volume Editor: Jordan, R.
Vol. 198, 2006

Surface-Initiated Polymerization I
Volume Editor: Jordan, R.
Vol. 197, 2006

Conformation-Dependent Design of Sequences in Copolymers II
Volume Editor: Khokhlov, A. R.
Vol. 196, 2006

Conformation-Dependent Design of Sequences in Copolymers I
Volume Editor: Khokhlov, A. R.
Vol. 195, 2006

Enzyme-Catalyzed Synthesis of Polymers
Volume Editors: Kobayashi, S., Ritter, H., Kaplan, D.
Vol. 194, 2006

Polymer Therapeutics II
Polymers as Drugs, Conjugates and Gene Delivery Systems
Volume Editors: Satchi-Fainaro, R., Duncan, R.
Vol. 193, 2006

Polymer Therapeutics I
Polymers as Drugs, Conjugates and Gene Delivery Systems
Volume Editors: Satchi-Fainaro, R., Duncan, R.
Vol. 192, 2006

Interphases and Mesophases in Polymer Crystallization III
Volume Editor: Allegra, G.
Vol. 191, 2005

Wax Crystal Control · Nanocomposites Stimuli-Responsive Polymers

With contributions by
S. Aoshima · F. R. Costa · L. J. Fetters · G. Heinrich · S. Kanaoka
A. Radulescu · D. Richter · M. Saphiannikova · U. Wagenknecht

Springer

The series *Advances in Polymer Science* presents critical reviews of the present and future trends in polymer and biopolymer science including chemistry, physical chemistry, physics and material science. It is adressed to all scientists at universities and in industry who wish to keep abreast of advances in the topics covered.
As a rule, contributions are specially commissioned. The editors and publishers will, however, always be pleased to receive suggestions and supplementary information. Papers are accepted for *Advances in Polymer Science* in English.
In references *Advances in Polymer Science* is abbreviated *Adv Polym Sci* and is cited as a journal.

Springer WWW home page: springer.com
Visit the APS content at springerlink.com

ISSN 0065-3195
ISBN 978-3-642-09474-3 e-ISBN 978-3-540-75500-5
DOI 10.1007/978-3-540-75500-5

Springer is a part of Springer Science+Business Media

springer.com

Softcover reprint of the hardcover 1st edition 2008

Cover design: WMXDesign GmbH, Heidelberg

Printed on acid-free paper 02/3180 YL – 5 4 3 2 1 0

Editorial Board

Advances in Polymer Science Also Available Electronically

For all customers who have a standing order to Advances in Polymer Science, we offer the electronic version via SpringerLink free of charge. Please contact your librarian who can receive a password or free access to the full articles by registering at:

springerlink.com

If you do not have a subscription, you can still view the tables of contents of the volumes and the abstract of each article by going to the SpringerLink Homepage, clicking on "Browse by Online Libraries", then "Chemical Sciences", and finally choose Advances in Polymer Science.

You will find information about the

- Editorial Board
- Aims and Scope
- Instructions for Authors
- Sample Contribution

at springer.com using the search function.

Contents

Adv Polym Sci (2008) 210: 1–100
DOI 10.1007/12_2007_124

Published online: 19 October 2007

Polymer-Driven Wax Crystal Control Using Partially Crystalline Polymeric Materials

A. Radulescu[1,2] · L. J. Fetters[3] · D. Richter[1] (✉)

[1]Institut für Festkörperforschung, Forschungszentrum Jülich, 52425 Jülich, Germany
d.richter@fz-juelich.de

[2]*Present address:*
JCNS-FRM II, Lichtenbergstr. 1, 85747 Garching b. München, Germany

[3]Department of Chemical and Biomedical Engineering, Cornell University,
Ithaca, NY 14853-5021, USA

Abstract A long term problem confronting the transportation of crude oils and refined middle distillate fuels is the abrupt degradation of the system viscoelastic properties as temperatures fall below ~0 °C. The prime contributor to this unfavorable event is the phase separation of paraffins (waxes) with carbon contents ranging from C_{16} to ~C_{38}. This problem has been addressed, with varying degrees of success, via the use of formulations containing polymeric additives. Additives that have had long use are the ethylene-rich copolymers of ethylene and vinyl acetate (EVA). Although far from being universally successful in their treatment capacity, the EVA materials serve as a prototype wax-crystal modifier in that their structure of alternating amorphous-crystalline segments serves as a model for regarding the composition of other polymeric candidates. A recent candidate is the diblock copolymer consisting of ethylene and ethylene-butene segments. This material thus consists of a semicrystalline block joined to an amorphous counterpart. After a four-year development period it became a commercial item in 2000. In hydrocarbon milieu the polyethylene block will self-assemble as the system temperature decreases to yield plate-like micelles that remain in solution due to the presence of the amorphous "hairs". Small angle neutron scattering studies have shown that this polymer architecture is quite effective in providing a scaffold for wax nucleation, thus leading to quite effective control of wax crystal size in a variety of fuels. An architectural mimic of EVA is the random copolymer of ethylene and butene. This particular random copolymer was also shown to be highly effective in its capacity as a modifier for wax crystal size control. The mechanism by which this is done was found to be even richer than that shown by the diblock architecture.

Keywords Partially crystalline copolymers · Structure and morphology via SANS · Wax crystal modification · Yield stress behavior

Abbreviations

1,4 PBd	1,4 Deuterated polybutadiene isomer
1.5HH5DH	PE-PEP diblock copolymer with 1.5 K PE block obtained from hydrogenated butadiene and saturated by hydrogen and 5 K PEP block obtained from deuterated butadiene and saturated by hydrogen
5DH/8HH	PE-PEP diblock copolymer with 5 K PE block obtained from deuterated butadiene and saturated by hydrogen and 8 K PEP block obtained from hydrogenated butadiene and saturated by hydrogen
6HH/10DH	PE-PEP diblock copolymer with 6 K PE block obtained from hydrogenated butadiene and saturated by hydrogen and 10 K PEP block obtained from deuterated butadiene and saturated by hydrogen
6HH/15DH	PE-PEP diblock copolymer with 6 K PE block obtained from hydrogenated butadiene and saturated by hydrogen and 15 K PEP block obtained from deuterated butadiene and saturated by hydrogen
APSA	Available platelet surface area
C_{24}	Tetracosane paraffin wax
C_{28}	Octacosane paraffin wax
C_{30}	Triacontane paraffin wax
C_{32}	Dotriacontane paraffin wax
C_{36}	Hexatriacontane paraffin wax
CCl_4	Carbon tetrachloride
CFPP	Cold filter plugging point
C_n	Normal paraffin (C_nH_{2n+2})
CP	Cloud point
DSC	Differential scanning calorimetry

EVA	Ethylene/vinylacetate
EVA-ga	EVA "growth arrestor"
EVA-nu	EVA "nucleator"
EVA-ss	EVA "single shot"
FSANS	Focusing small angle neutron scattering
H/D	Hydrogen/deuterium substitution
ILL	Institut Laue Langevin
JCNS	Jülich Center for Neutron Science
MDFI	Middle distillate fuel improvers
PB	Polybutadiene
PB-PI	Polybutadiene-polyisoprene
PE	Polyethylene
PEB-*n*	Poly(ethylene-butene) random copolymers with *n* the number of ethyl branches per 100 backbone carbons
PEP	Ethylene-propylene copolymer
PE-PEP	Polyethylene-(*alt*-ethylene-*co*-propylene) diblock copolymer
PP	Pour point
ppm	Parts per million
SANS	Small angle neutron scattering
TEM	Transmission electron microscopy
USANS	Ultra-small angle neutron scattering

List of symbols

b_i^{M}	Scattering length of the atom *i* in component *M*
b_j^{S}	Scattering length of the atom *j* in component *S*
$S_{\mathrm{bb}}(Q)$	Partial scattering function for fluctuation or blob scattering
$\sum w_n(D)$	One dimensional Patterson function
V_0^{PE}	Volume of a monomer in the PE subchain
v_0^{PEP}	Volume of a PEP monomer
m_0^{PEP}	Monomer mass
ζ_{PEP}	PEP density
$\Phi_{\mathrm{pol}}^{\mathrm{agg}}$	Volume fraction of the polymer in the aggregate
$\Phi_{\mathrm{pol}}^{\mathrm{sol}}$	Volume fraction of the polymer in solution
$\Phi_{\mathrm{wax}}^{\mathrm{layer}}$	Wax concentration within the layer
$\Phi_{\mathrm{pol}}^{\mathrm{layer}}$	Polymer concentration within the layer
$\Phi_{\mathrm{wax}}^{\mathrm{plate}}$	Volume fraction of wax platelets
$\Phi_{\mathrm{pol}}^{\mathrm{plate}}$	Volume fraction of polymer platelets
λ	Neutron wavelength
p	Power-law exponent
ν	Flory exponent
Γ	Gamma function
Φ	Total polymer volume fraction
α	Angle between axis of symmetry of a cylinder and the scattering vector Q
ξ	Mean blob size
Ω	Surface density of the blobs
l	Length of a monomer
δ	Geometrical factor somewhat larger than one
$\nu = 1/2$	Flory exponent for Gaussian chains
$\nu = 3/5$	Flory exponent for swollen coils

$\Delta\rho$	Scattering contrast factor
$\Delta\Omega$	Solid angle covered by detector element
$\Delta\rho(\underline{r})$	Scattering length density distribution
$\Delta\rho_b$	Scattering length contrast between the brush and the solvent
$\Delta\rho_c$	Scattering length contrast between the core and the solvent
$\Phi(z)$	Lateral platelet profile
Φ_{agg}	Volume fraction of the aggregates
σ_b	Gaussian width describing the brush
$\Phi_B(z)$	Polymer density within the brush along z-direction
γ_c	Critical angle of a neutron guide
σ_c	Gaussian width describing the core
$\Phi_C(z)$	Polymer density within the core along z-direction
ϕ_{cryst}	Volume fraction of the crystalline diblock
ζ_{cryst}	Density of the crystalline aggregate
σ_D	Smearing of the interplatelet distance D
ρ_i	Scattering length density of the component i
Σ_i	Component i of measured intensities vector
$\phi_i(r)$	Spatially varying density profile of the component i
ζ_{PE}	Polyethylene density
δQ	Uncertainty of the momentum transfer
ρ_s	Scattering length density of the solvent
ρ_{si}	Scattering length density of component i and type s
Φ_{wax}	Wax volume fraction
$[F^{\circ}]_{def}$	Defect energy for the incorporation of all ethyl groups into a large PE crystal
^{13}C NMR	Carbon 13 NMR
^{1}H NMR	Proton NMR
2θ	Scattering angle
a	Radius of cylinder
$A(Q, R, L, \alpha)$	Form factor for an ensemble of isotropically oriented cylinders
A, B and C	Terms of the quadratic regression problem that relates the measured intensities to the scattering length densities
A_2	Second virial coefficient
A-B	Block-copolymer
A_{wax}/V	Platelet area per cubic centimeter
B	Power-law prefactor
b_C	Scattering length of carbon
b_D	Scattering length of deuterium
b_H	Scattering length of hydrogen
c	Average amplitude of the modulation
C_1	Prefactor of the entropic contribution to the free energy
C_2	Prefactor of the enthalpic contribution to the free energy
$C_p(z)$	Perpendicular rectangular density profile
d	Core thickness, plate thickness
D	Stacking period in real space
$\mathrm{d}\sigma/\mathrm{d}\Omega$	Microscopic scattering cross-section
$D(u)$	Dawson function
D^*	Fractal object of dimension D^*
d_{eff}	Effective thickness of platelets
d_{pol}	Thickness of polymer layer

d_{wax}	Thickness of wax layer
E_{def}	Defect energy
E_{f}	Folding energy
f_{arm}	Number of arms of a star polymer
F_{core}	Enthalpic contribution to the free energy as a consequence of chain folding
F_{def}	Defect contribution to the free energy
F_{rod}	Perpendicular area of the rod
G	Classical Guinier prefactor
g	Number of monomers in the blob
h	Period of density modulation
$I(0)$	Forward scattering
$I(Q)_{\mathrm{exc}}$	Excess brush scattering
I_0	Incoming neutron flux
I_{ex}	Contribution from the concentration fluctuation in the brush
J_1	Cylindrical Bessel function of first order
k_{B}	Boltzmann's constant
$L = 2N_a$	Cylinder length
L_{b}	Brush length
Lu	Luminosity of the source
L_x	Dimension of a rectangular tile in direction x
L_y	Dimension of a rectangular tile in direction y
M_{cryst}	Molecular weight of the crystalline part of the block copolymer
M_{PE}	Molecular weight of the polyethylene chains
m_{s}	Geometrical factor describing the fraction of amorphous polymer in the brush
M_{w}	Molecular weight
N	Number of scattering objects
N_A	Avogadro number
n_{b}	Number of blobs per chain
n_{f}	Number of folds per chain
N_{p}	Number of platelets
N_{PE}	Chain length of the PE subchain
N_{PEP}	Number of monomers in the PEP segment
n_{s}	Number of methyl units in the neighborhood of the fold
$P(Q)$	Form factor
P_4	Porod constant
P_{rs}	Partial scattering functions
Q	Momentum transfer
Q_z	z-Axis in reciprocal space
$\underline{r}$	Position vector
R	Disk radius (platelet lateral extension)
$R(a, Q)$	Scattering amplitude
$r_1 r_2$	Reactivity ratio product
R_{g}	Radius of gyration
RV	Corresponding reciprocal lattice vector
S_ν	Surface
$S(Q)$	Structure factor
SA	Surface area per chain
$S_{ij}(Q)$	Partial structure factor
$S_N(Q)$	Structure factor of N stacking platelets

T	Sample transmission factor
t-$(CH_3)_3COK$	Tertiary-butyl potassium alkoxide
V_0	Volume of one scattering object
V_{arm}	Molar volume of one arm of a star polymer
v_M	Effective volume occupied by the component M
V_p	Molar polymer volume
v_S	Effective volume occupied by the component S
V_{sample}	Volume of the sample (volume of dissolved polymer in case of solutions)
Σ CFPP	Sum of cold filter plugging point (CFPP) results
$\Delta\Phi$	Difference between the aggregated and free polymer volume fractions
$\Delta\rho_{plate}$	Contrast between the solvent and the platelet
$\Delta\rho_{pol}$	Polymer scattering contrast
$\Delta\rho_{rod}$	Contrast between the solvent and the rod
$\Delta\rho_{wax}$	Wax scattering contrast
Φ_{plate}	Volume fraction of platelets
Φ_{rod}	Volume fraction of rods

1 Introduction

Crude oils and refined middle distillate products such as diesel fuel, kerosene (jet fuel), or heating oil contain an important fraction of paraffins of high energy content with a broad linear and branched chain length distribution ($\sim C_{16}$ *to* $\sim C_{38}$) [1]. Depending upon the type of crude deposits and the refining technology applied, this fraction can vary between 10 and 30 wt % [2, 3]. A typical paraffinic profile for a diesel fuel [3] is given in Fig. 1. These complex fluids can undergo a dramatic change in the viscoelastic properties upon cooling below ambient temperature because of the precipitation of the long chain paraffins (or waxes) in plate-like crystals with sizes of hundreds of micrometers and an overall morphology resembling a "house-of-cards" [4]. An example of this morphology is shown in Fig. 2.

For example, the C_{36} wax forms stable crystals that interlock to form a solid network. This network traps a large quantity of solvent (in the case presented in Fig. 2 the solvent is 1-octanol). The wax molecules that crystallize in the lamellar orthorhombic phase are fully extended ($\sim$47 Å). Liquid state molecules do not occupy interlamellar sites. The packing of C_{36} in the gel is the same as in the neat orthorhombic phase. In wax mixtures, a single packing layer has a thickness equal to the average length of the wax molecules. A wax molecule with a carbon number greater than the average carbon number bends to insert itself inside the layer and associate itself with a molecule with a carbon number lower than the average. The mismatch between the length of the molecules and the thickness of the packing layer causes conformational disorders in the interlamellar regions of the crystals but this disorder imparts to these crystals their wax-like properties.

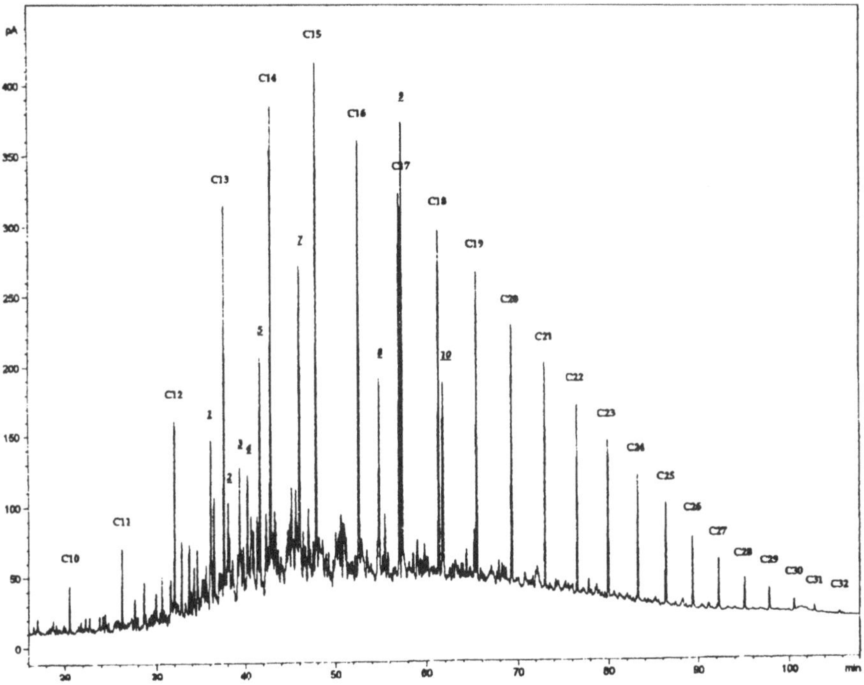

Fig. 1 Example of a chromatographic analysis of a diesel fuel: the *n*-paraffins are identified by their chain length

This aggregation event results in a reduction of the ease of flow of crude oils and a loss of fluidity and filterability of middle distillates. This can cause technical problems related to pump, pipeline, and winter time diesel engine filter plugging. It also causes problems in the case of crude oils extracted from deep-sea reservoirs with pipeline transport through sub –0 °C regions. Paraffin deposition (see Fig. 3 [5]) can also occur in storage tanks and conduits. Removal of these waxy paraffins on a large scale is expensive and also counter-productive since that exercise perversely can eliminate an important fraction of the high-density paraffins (high-energy fraction) from the fuel.

The pour point (PP) – representing the temperature at which the system gels and becomes mechanically rigid [2] – is about 10 °C for a typical untreated crude oil [1] while, in the case of untreated diesel fuels it may vary over a wide temperature range below 0 °C [2]. A parameter that directly correlates to the occurrence of the low temperature technical problems caused by the wax crystals is the cold filter plugging point (CFPP). This parameter corresponds to the temperature when plugging occurs in a 45 μm filter under standardized conditions. For untreated diesel fuels, the CFPP temperature is normally a few degrees higher than the PP [2]. In order to circumvent these technical problems chemical additives come into play and can be classified as middle distillate fuel

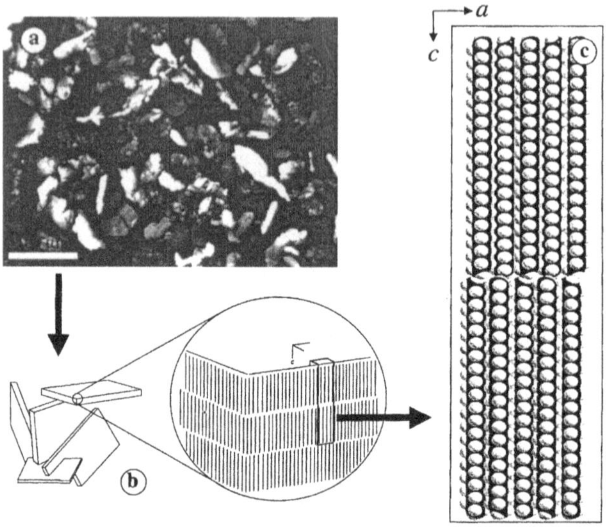

Fig. 2 Optical micrograph of a 4% C_{36}/1-octanol gel (**a**), a cartoon representation of the wax platelets forming "house-of-cards" morphology (**b**) and the molecular packing of C_{36} within the one-dimensional multilayer wax crystal. The *scale bar* in **a** represents 100 μm

Fig. 3 Plugged pipeline

improvers (MDFI). These polymeric species are random, graft, or block copolymers (containing amorphous and crystalline segments) that can self-assemble in solution to yield stable micelles. These micelles can serve as templates for the wax crystals as the system temperature passes through that of the PP. Historically, commercial formulations containing the ethylene-rich copolymers of ethylene and vinyl acetate (EVA) have been available from ESSO. Those additives that modify the size and shape of the wax crystals appearing in a fuel when cooled below its cloud point (CP) – the temperature when the wax starts to crystallize – are also called "pour-point depressants" [2, 6] (where CP > PP). Their main role is to maintain the wax crystals at moderate sizes and to prevent their association into larger structures. Polymeric materials that qualify as wax crystal modifiers share a common structural characteristic with the wax population. This pertains to the joint presence, at room temperature, of crystalline and amorphous segments. This allows, in dilute solution, the self-assembly of the crystalline segments as temperature decreases, which in turn can provide stable platforms that serve as sites for wax crystal formation. These micelles, as they take on wax, are stabilized in solution via the amorphous fraction of the parent copolymer.

This self-assembly behavior of certain crystalline-amorphous copolymers (and their interactions with wax) has been studied in recent years [7–18] via the application of small-angle neutron scattering (SANS) techniques. This encompassed the classical pin-hole SANS, the focusing SANS (FSANS), and the ultra-SANS (USANS) double-crystal diffractometry. This allowed the qualitative and quantitative analysis of the morphologies formed over the wide length scale from nanometer to micrometer sizes. By applying contrast-matching techniques, separate analyses of the wax and copolymer conformations within the common morphologies were studied. The visibility of the molecules in solution depends upon the difference between the solvent/solute scattering length densities. Of the available scattering techniques, only SANS has this required versatility via contrast control needed for the detailed analysis of the paraffin/polyolefin self-assembled structures. The advantage of neutrons over X-rays is the ability to vary the scattering length densities of different constituents of a hydrocarbon sample over a broad range by hydrogen/deuterium (H/D) substitution.

The overall morphology of the wax copolymer aggregates at the micrometric scale was also investigated by optical and transmission electron microscopy as complementary techniques for the USANS investigations. The results obtained allowed the clarification of the self-assembled structures in solution of the crystalline-amorphous random copolymers. This has led to a broader understanding of wax crystal modification and control in solution than previously available. As noted, SANS is the required technique to study these hydrocarbon mixtures since contrast is easily achieved via the judicious mixture of hydrogenated and deuterated materials. As detailed experimentation has shown [7–18], this experimental capacity is vital for a complete survey of the structures formed via the interactions of the paraffinic wax with the scaffolds formed by the self-assembling chains. The self-assembled micelles thus identified have proven to provide highly efficient structures for controlled wax nucleation in refined middle distillate fuels.

1.1 Self-Assembling Copolymers

Amorphous diblock copolymers in selective solvents form micelles, a phenomenon similar to that shown by the amphiphilic surfactants. Many groups have studied the micellar behavior in polymer systems where the micellization is based on the incompatibility between one of the polymer components and the solvent [19–21]. Such experiments have been performed with a variety of techniques such as low angle and dynamic laser light scattering [22–25], viscometry [26, 27], SANS [28], and small angle X-ray scattering [29, 30]. The micelles often display a core-shell geometry in which the insoluble block forms the inner part or core whereas the soluble block forms a solvent-rich brushy shell. Generally, the micellization process is driven by the systems tendency to lower the interfacial free energy by minimizing the area of the insoluble block exposed to the solvent. On the other hand, according to scaling theories, the micellar growth is mainly counterbalanced by chain stretching of the core and shell that increases the free energy of micellization. The relative contributions of core and shell chain stretching to the free energy depend on the core and shell dimensions, primarily determined by block copolymer molar mass characteristics.

The copolymers of vinylacetate and ethylene (EVA) have a long commercial history, for example, in adhesive formulations. The copolymerization of these monomers can be classified [31] as almost ideal since $r_1 r_2 \approx 1$. Here r_1 and r_2 are the reactivity ratios of ethylene and VA, respectively ($r_1 = 0.8$ and $r_2 = 1.2$). Hence, to a first approximation the monomer incorporation can be random. In spite of these congenial reactivity ratios the EVA materials show characteristics indicative of severe compositional variations; see below.

The materials rich in ethylene have been used as wax crystal modifiers. Although useful, these materials have proven to be inadequate in certain MDFI applications, and in extreme cases virtually devoid of activity. As will be discussed later, the EVA family contains fractions of high ethylene content copolymer that cause precipitation (and thus a loss in activity). For example, the commercial EVA "nucleator" grade (Infinium) sacrifices ~55 wt % of the added copolymer [18] to premature phase separation in hydrocarbons. This is a consequence of the communal compositional unruliness inherited from the commercial high-pressure, high-temperature free-radical polymerization process. Another potential loss would occur when the VA content increases to ~30 wt %. This composition can yield amorphous (and thus soluble) elastomers devoid of any self-assembling capacity.

This behavior appears to be a congenital defect for the EVA grades as they are currently produced. It is clear that the commercial designations of "nucleator" (EVA-nu), "growth arrestor" (EVA-ga) and "single shot" (EVA-ss) are heuristic and do not explain mechanistically the role of these polymers in modifying the wax crystal structure. The VA units in the commercial products are nominally present at ~9% (mole fraction) in the EVA-nu, ~12% in the EVA-ss, and ~15% in the EVA-ga. Each grade denotes the perceived function of each copolymer type. These commercial packages usually consist of EVA-ss or mixtures of the nu and ga grades. The ss grade is assumed to combine both the nucleator and growth arrestor characteristics found in the EVA mixture.

A noticeable improvement in formulation performance activity (relative to the EVA materials) was encountered when a crystalline-amorphous butadiene-based diblock was introduced in formulation packages in 2000. This polymer is prepared via the sequential anionic living polymerization of butadiene to give a block structure (1.5 kg/mol) of near 1.4 content coupled to a segment (5 kg/mol) containing a 1.2–1.4 mixed microstructure. The rationale for these molecular weights will be given later. Hydrogenation yields a semicrystalline-amorphous polyolefin diblock that forms "hairy" platelet micelles in hydrocarbon solution. This self-assembled structure serves, in turn, as an efficient site for wax formation. The polymerization uses *n*-butyllithium in toluene. The polymerization of the second segment is conducted with t-$(CH_3)_3COK$ present as the butadiene microstructure modifier [32]. The initial butadiene diblock microstructure yields, after hydrogenation, a semicrystalline segment followed by an amorphous, hydrocarbon soluble block that serves to stabilize the self-assembled miceller structure in the fuel.

Hydrogenated random copolymers of 1.4- and 1.2-butadiene represent another class of crystalline-amorphous materials [10–15, 17]. Following hydrogenation, the resultant copolyolefin contains ethylene (E) and 1-butene (B) units. These materials can be described as PEB-n where n denotes the number of ethyl units per 100 backbone carbons. For samples where $n \leq 12$ the samples will show partial crystallinity at room temperature while for $n \geq 13$ the samples are amorphous in the bulk state. The latter PEB copolymers lack self-assembly capacity. An equivalent amorphous segment is derived from polyisoprene. In this case hydrogenation yields the essentially alternating ethylene-propylene copolymer (PEP). This material also contains about 7% isopropyl units that randomly appear between the ethylene-propylene units.

For these copolymers in decane the driving force toward aggregation comes from the tendency of the PE component to crystallize as the system is cooled. The PE component crystallizes with high enthalpy gain and delivers the thermodynamic reason for aggregation. From the lamellar morphology of polyethylene crystals two-dimensional structures are also expected for these diblocks with a semicrystalline polyethylene core surrounded by a PEP or PEB-12 amorphous brush. The detailed morphology of such aggregates will be determined by the balance of entropic forces resulting from the stretching of the brush hairs and the enthalpic contribution from the polyethylene chain folding, as we will discuss in detail in Sect. 4.

It was concluded that control of the wax crystallization process involving the PE-PEP type of diblock is carried out by the nucleation of waxes via the crystallizable segments. This behavior invited the evaluation of PEB-based random copolymer architectures where semicrystalline and amorphous segments are combined in an alternative manner. One such family is represented by nearly random copolymers of ethylene and 1-butene obtained by the anionic polymerization of butadiene with variable 1.2- and 1.4- modes of addition [32]. ^{13}C NMR was used [33, 34] to evaluate the sequence distribution characteristics of the PEB materials. The results were cast in terms of the reactivity ratio products as outlined by Hsieh and Randall [35]. The mean $r_1 r_2$ value of 0.73 (±0.06) suggests essentially random character with, perhaps, a slight tendency toward alternation. The PEB materials are not, strictly speaking, random ethylene-butene copolymers since the minimum number of ethylene units between 1-butene structures occurs in multiples of two. This is the automatic consequence of the 1,4-units in the parent chain. The microcrystallinity of the PEB-*n* copolymers can be tuned by changing the ratio of ethylene to butene units.

2
Small Angle Neutron Scattering

Neutrons interact with matter via the strong interaction and hence see the nuclei rather than the diffuse electron cloud observed by X-rays. This has major advantages such as to be able to see light atoms in the presence of heavier ones and to distinguish neighboring elements more easily. Moreover, the cross-section of an atom generally varies between isotopes of the same element allowing the exploitation of isotopic substitution methods to highlight structural and dynamic details. In particular, the strong difference in cross-section between hydrogen and deuterium enables contrast variation methods in the investigation of synthetic macromolecules and biomaterials.

In this section, we will first lay out the general principles of SANS instruments, presenting both the conventional pin-hole camera as well as the novel focusing small angle scattering instrument. Thereafter, we will introduce SANS cross-sections, which may be obtained in absolute terms. We will then relate to relevant theoretical SANS form and structure factors. Starting from simple structures, we will address general expressions for fractal geometries and then discuss in some detail the cross-sections for platelets with internal structure and those for rods with longitudinal density modulation. Finally, we will briefly present structure factors for a paracrystalline order. All the form and structure factors discussed will be important for the further understanding of the experimental results and discussions presented later.

2.1 Small Angle Neutron Scattering Instruments

Elastic scattering with neutrons reveals structural information on the arrangement of atoms and magnetic moments in condensed matter. On a mesoscopic scale, such arrangements may be macromolecules, self-assembled systems such as polymeric micelles, biological membranes, proteins, or biomolecular aggregates. Scattering delivers information about the size, the number density, and the correlation between the objects. The information in such scattering experiments is contained in the neutron intensity measured as a function of the momentum transfer Q:

$$Q = \frac{4\pi}{\lambda} \sin\theta \,, \tag{1}$$

where λ is the neutron wavelength and 2θ the scattering angle. Scattering experiments explore matter in reciprocal space and Q acts as a kind of inverse yardstick. Large Q values relate to short distances while a small Q relates to large objects. Aiming for mesoscopic scales, SANS is optimized for the observation at small scattering angles and thus large objects. The principle layout of a conventional pin-hole SANS instruments is shown in Fig. 4. Such instruments are placed at cold neutron sources, which provide long neutron wavelengths resulting in small Q at a given scattering angle (see Eq. 1). In such an instrument the beam monochromatization is performed by a velocity selector in front of the instrument. A velocity selector is a rotor with spiral groves rotating at a speed up to 30 000 rpm. The spiral groves are manufactured such that the rotor becomes transparent for a given wavelength band depending on the speed of the rotor. The pitch of the spiral determines the width of the transmitted wavelength distribution. Typical velocity selectors provide wavelength resolutions of $\Delta\lambda/\lambda \cong 10$–20%.

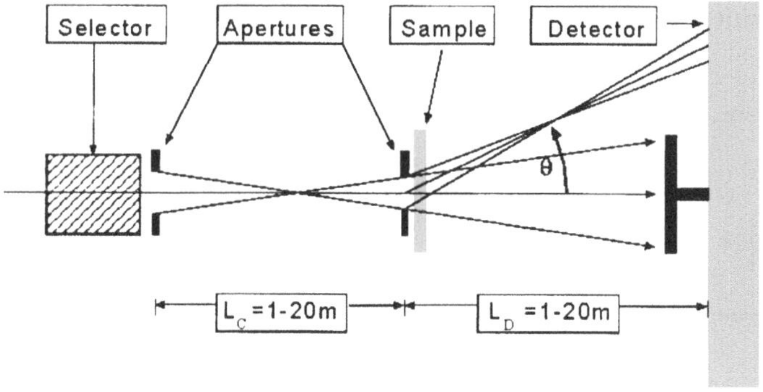

Fig. 4 Principle of pin-hole SANS

After the velocity selector, a collimation section determines the divergence of the incoming neutron beam. This divergence has to be optimized according to the angular range being investigated. The sample is placed after the collimation section. The scattered neutrons are counted on a two-dimensional position-sensitive detector, which today covers typically an area of 1 m^2. The non-scattered neutrons in the primary beam are absorbed at a beam stop in front of the detector. For a given instrumental setting, neutrons may be detected only in a limited angular interval. Therefore, the distance between detector and sample may be varied between about 1 m and 40 m (D11 at the ILL in Grenoble). Using neutron wavelengths between 5 and 15 Å the Q interval between 5×10^{-4} Å^{-1} and 0.3 Å^{-1} becomes accessible. An intensity-optimized setting of the instrument is achieved if the collimation section equals the sample detector distance. Thus, the length of the collimation section has to be adjusted to the sample detector distance. For a geometrically optimized machine, the intensity at the sample position is given by:

$$I_0 \simeq Lu \left(\frac{\delta Q}{k}\right)^4 L_D^2 . \qquad (2)$$

Here Lu is the luminosity of the source, δQ the uncertainty of the momentum transfer of the scattered beam (defining also the smallest accessible Q value), and $k = 2\pi/\lambda$ the wavevector of the incoming neutrons. Equation 2 shows that the intensity at the sample is proportional to the square of the length of the machine. Therefore, SANS instruments have a typical length of 40 m reaching 80 m for D11 at the ILL.

In order to increase further the Q resolution of a SANS instrument towards smaller values, focusing optical elements need to be introduced. One possibility are neutron lenses placed in front of the sample such that the incoming neutron beam is focused on the detector [36, 37]. Due to the very small index of refraction a large number of lenses are necessary. At present intense efforts are underway in order to provide lenses with sufficient surface quality in order to suppress surface scattering.

The principle of a fully focusing SANS [38–40] is presented in Fig. 5.

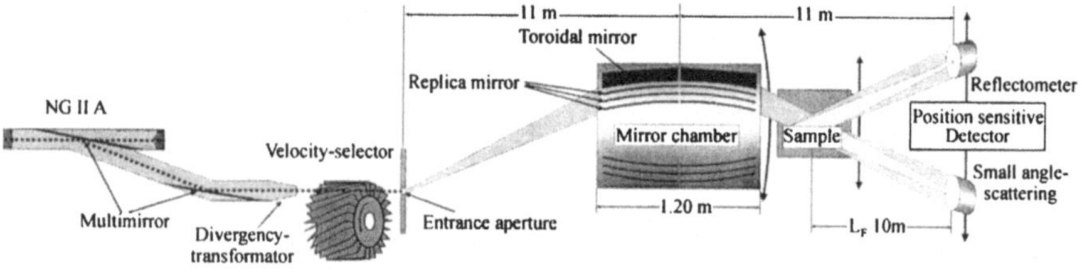

Fig. 5 Schematic layout of the focusing SANS camera of the Jülich Center for Neutron Science (JCNS) at the FRM-II reactor in Munich

In such an instrument the monochromated neutrons enter through an aperture with a diameter of about 1 mm and then hit the focusing mirror with the full divergence provided by the neutron guide. From the mirror, the neutrons are reflected and focused to a position-sensitive detector in the focal plane. Just behind the mirror the sample is positioned. Figure 6 presents a view on the focusing mirror used in the Jülich Centre for Neutron Science (JCNS) SANS instrument.

Fig. 6 View of the focusing mirror of the JCNS focusing SANS instrument

Just behind the mirror the sample is positioned. With a focusing SANS instrument of a total length of 20 m with $\lambda = 12\,\text{Å}$ neutrons a Q range 10^{-3}–$10^{-4}\,\text{Å}^{-1}$ is accessed extending the pin-hole SANS Q range by another order of magnitude. For such an instrument the intensity at the sample position is given by:

$$I_0 \simeq L(4\gamma_c)^2 \left(\frac{\delta Q}{k}\right)^2 L_D^2 . \tag{3}$$

For small δQ, this instrument is superior to the conventional pin-hole instrument Eq. 2 because neutrons from the full solid angle provided by the neutron guide ($4\gamma_c$), where γ_c is a critical angle of the neutron guide, can be used [41]. Though the concept of the focusing SANS was developed some time ago [42], it could only be constructed recently because of the very high demand on the surface quality of the focusing mirror. Such mirrors today may be manufactured as a result of an extended development project regarding X-ray telescopes for satellites [43, 44].

2.2 SANS Cross-sections

In the field of synthetic macromolecules and generally soft matter, the advantage of neutrons over X-rays is the ability to vary the scattering contrasts between different constituents of a hydrocarbon sample over a broad range by H/D substitution. Since the molecules affected by H/D exchange are chemically the same, the physical chemistry of the samples is only marginally modified, if at all. The visibility of an object in solution for example depends on the difference of the solvent/solute scattering length densities:

$$\Delta\rho_{\mathrm{M}} = (\rho_{\mathrm{M}} - \rho_{\mathrm{S}}) = \frac{\sum_{i=1}^{\text{atoms in M}} b_i^M}{v_{\mathrm{M}}} - \frac{\sum_{j=1}^{\text{atoms in S}} b_j^S}{v_{\mathrm{S}}} \tag{4}$$

where b_i^M denotes the scattering length of the different atoms in M and b_j^S those of the atoms in S. v_{M} and v_{S} are the effective volumes occupied by the objects composed of the atoms in the respective sums. M may denote a monomer segment of a polymer and S a solvent molecule. Since $b_{\mathrm{H}} = -3.57 \times 10^{-13}$ cm, $b_{\mathrm{D}} = 6.57 \times 10^{-13}$ cm, and $b_{\mathrm{C}} = +6.65 \times 10^{-13}$ cm H/D replacement allows for huge variations of $\Delta\rho$ for hydrocarbons.

Using a mixture of suitable amounts of deuterated and protonated solvent molecules, in general it is possible to achieve zero contrast, $\Delta\rho_{\mathrm{M}} = 0$, for one component M of the system. This technique of contrast matching and variation is of essential importance for the investigation of complex multi-component systems such as A–B block-copolymer aggregates and the associated wax molecules.

The contribution to the scattering intensity from some object described by its scattering length density distribution $\Delta\rho(\underline{r})$ is:

$$I(\theta) = TI_0 \frac{\mathrm{d}\sigma}{\mathrm{d}\Omega}\Delta\Omega \quad \text{and} \quad \frac{\mathrm{d}\sigma}{\mathrm{d}\Omega} = \left| \int \Delta\rho(\underline{r}) \mathrm{e}^{\mathrm{i}Qr}\, \mathrm{d}^3\underline{r} \right|^2 , \tag{5}$$

where $I(\theta)$ is the intensity scattered into one detector element, $\Delta\Omega$ the solid angle covered by that element, θ the scattering angle, I_0 the incoming neutron flux, and T the sample transmission factor. The structural information on the sample is contained in $\mathrm{d}\sigma/\mathrm{d}\Omega$. It is the absolute square of the Fourier transform of the scattering length density distribution $\Delta\rho(\underline{r})$. The squaring and the orientational averaging necessary in the case of anisotropic objects in an isotropic solution prevents a direct calculation of $\Delta\rho(\underline{r})$ from $I(\theta)$ by a simple inverse Fourier transform. For exact structural details one normally relies on model fittings (see later). Nevertheless, some general information on the structure may be obtained from the forward scattering $I(0)$ and the exponent p provided by $\mathrm{d}\sigma/\mathrm{d}\Omega \approx Q^p$ in different Q regimes.

With neutrons it is comparatively easy to obtain the scattering data on an absolute scale normalized to the sample volume. This is done normally by a comparison with scatterers where the absolute cross-section is known. For solutions, V_{sample} is taken as the volume of the dissolved polymer:

$$\frac{\mathrm{d}\Sigma}{\mathrm{d}\Omega} = V_{\text{sample}}^{-1} \frac{\mathrm{d}\sigma}{\mathrm{d}\Omega} . \tag{6}$$

We first consider the cross-section for one object where we have:

$$\frac{\mathrm{d}\sigma_0}{\mathrm{d}\Omega}(Q \to 0) = \left| \int \Delta\rho(\underline{r})\,\mathrm{d}^3 r \right| = \Delta\rho^2 V_0^2 . \tag{7}$$

Here V_0 is the volume of the object. Furthermore we have assumed that the object is homogeneously decorated with a constant contrast factor, $\Delta\rho$. If N of these objects are located at random positions and random orientations in a solution, then all interferences between the objects average out to zero and their scattering intensities add up to $N\,\mathrm{d}\sigma_0/\mathrm{d}\Omega$. A given polymer volume V in solution that splits into N aggregates yields $V_0 = V/N$. With that we have:

$$\frac{\mathrm{d}\Sigma}{\mathrm{d}\Omega}(Q \to 0) = \frac{1}{V}\frac{\mathrm{d}\sigma}{\mathrm{d}\Omega} = \frac{N}{V}(\Delta\rho)^2 \left(\frac{V}{N}\right)^2 = \Delta\rho^2 V_0 , \tag{8}$$

i.e., the forward scattering is given by the contrast factor times the scattering volume. This is a very simple and important result.

As alluded to above, the small angle scattering of neutrons arises from the fluctuations of the scattering length densities within a material. We now generalize to multi-component systems. In this case and under the assumption of incompressibility, the macroscopic cross-section is given by:

$$\frac{\mathrm{d}\Sigma}{\mathrm{d}\Omega}(Q) = \sum_{ij} (\rho_i - \rho_\mathrm{s})(\rho_j - \rho_\mathrm{s}) S_{ij}(Q) . \tag{9}$$

Here we have used the solvent with a scattering length density ρ_s as a reference. ρ_i represent the scattering length densities of the component i. The partial structure factor $S_{ij}(Q)$ is then defined as follows:

$$S_{ij}(Q) = \frac{1}{V} \int \left\langle \phi_i(\underline{r})\phi_j(\underline{r}') \exp[\mathrm{i}\underline{Q}(\underline{r} - \underline{r}')]\,\mathrm{d}^3\underline{r}\,\mathrm{d}\underline{r}' \right\rangle . \tag{10}$$

Integration is performed over the sample volume and $\phi_i(r)$ describes the volume fraction of monomer or molecule i at position r. For a ternary system containing oil, polymer, and paraffin with oil as a reference (solvents), the macroscopic cross-section in terms of partial structure factors then assumes the form:

$$\frac{\mathrm{d}\Sigma}{\mathrm{d}\Omega}(Q) = (\rho_\mathrm{p} - \rho_\mathrm{s})^2 S_\mathrm{pp}(Q) + 2(\rho_\mathrm{p} - \rho_\mathrm{s})(\rho_\mathrm{w} - \rho_\mathrm{s}) S_\mathrm{pw}(Q) + (\rho_\mathrm{w} - \rho_\mathrm{s})^2 S_\mathrm{ww}(Q) \,. \tag{11}$$

Here the subscripts p and w indicate polymer and wax (paraffin), respectively.

The partial scattering functions contain structural information about the polymer and wax aggregates and their mutual interaction. They in principle may be accessed separately by proper variation of the solvent scattering length density ρ_s, thereby varying the contrast factors in Eq. 11. When an experiment on an identical polymer wax system is performed using M different isotope mixtures of the solvent, Eq. 11 yields a set of linear equations relating the measured intensities, the scattering contrasts, and the partial scattering functions:

$$\underline{\Sigma} = \underline{\underline{M}}\,\underline{S} \,. \tag{12}$$

With the contrast matrix:

$$\underline{\underline{M}} = \begin{pmatrix} \Delta\rho_{\mathrm{p}_1}^2 & 2\Delta\rho_{\mathrm{p}_1}\Delta\rho_{\mathrm{w}_1} & \Delta\rho_{\mathrm{w}_1}^2 \\ \vdots & \vdots & \vdots \\ \Delta\rho_{\mathrm{p}_n}^2 & 2\Delta\rho_{\mathrm{p}_n}\Delta\rho_{\mathrm{w}_n} & \Delta\rho_{\mathrm{w}_n}^2 \end{pmatrix} \tag{13}$$

and the vectors:

$$\underline{\Sigma} = \begin{pmatrix} \left.\frac{\mathrm{d}\Sigma}{\mathrm{d}\Omega}\right|_1 \\ \vdots \\ \left.\frac{\mathrm{d}\Sigma}{\mathrm{d}\Omega}\right|_n \end{pmatrix} , \quad \underline{S} = \begin{pmatrix} S_\mathrm{pp}(Q) \\ S_\mathrm{pw}(Q) \\ S_\mathrm{ww}(Q) \end{pmatrix} . \tag{14}$$

Equation 12 may be regrouped in terms of the known quantities ρ_{si} and Σ_i yielding a quadratic regression problem:

$$A - B\rho_{si} + C\rho_{si}^2 = \left(\frac{\mathrm{d}\Sigma}{\mathrm{d}\Omega}\right)_i , \tag{15}$$

with:

$$\begin{aligned} A(Q) &= \rho_\mathrm{p}^2 S_\mathrm{pp}(Q) + 2\rho_\mathrm{p}\rho_\mathrm{w} S_\mathrm{pw}(Q) + \rho_\mathrm{w}^2 S_\mathrm{ww}(Q) \\ B(Q) &= 2\rho_\mathrm{p} S_\mathrm{pp}(Q) + 2(\rho_\mathrm{p} + \rho_\mathrm{w}) S_\mathrm{pw}(Q) + 2\rho_\mathrm{w} S_\mathrm{ww}(Q) \\ C(Q) &= S_\mathrm{pp}(Q) + 2S_\mathrm{pw}(Q) + S_\mathrm{ww}(Q) \,. \end{aligned} \tag{16}$$

From the quadratic regression analysis the best values for A, B, and C can be obtained for each Q. For this purpose a Q-dependent least square analysis needs to be performed. Equating 16 with respect to the partial structure factors, the desired scattering functions may be calculated.

2.3 Small Angle Scattering from Simple Structures

At high temperatures the polymers and paraffins are completely dissolved in decane and a small angle scattering experiment will provide the scattering from single coils. For this case the corresponding macroscopic cross-section exhibits the following form:

$$\frac{\Delta\rho^2}{\frac{d\Sigma}{d\Omega}(Q)} = \frac{1}{\phi V_p P(Q)} + 2A_2 \,. \tag{17}$$

Here A_2 denotes the second viral coefficient, $P(Q)$ the polymer form factor, and V_p the molar polymer volume. For a Gaussian coil:

$$P(Q) = \frac{2}{\left(Q^2R_g^2\right)^2}\left(e^{-Q^2R_g^2} - 1 + Q^2R_g^2\right), \tag{18}$$

is the famous Debye function [45], where R_g is the radius of gyration. The form factor of a swollen coil may be expressed in terms of the Beaucage function (see later).

As the temperature is lowered the self-assembling of the polymers and paraffins results in the formation of large objects with a characteristic morphology for each investigated component. The scattering from these objects can often be characterized by the power laws:

$$\frac{d\Sigma}{d\Omega}(Q) \sim Q^{-p} \,, \tag{19}$$

where the exponent p signifies the spatial arrangements of the polymers and paraffins [28, 46]. For single chains p is equal to $1/\nu$, and the Flory exponent assumes $\nu = 3/5$ for swollen coils and $\nu = 1/2$ for Gaussian chains. For objects of general fractal structure, p denotes the fractal dimension D^* of the object. If the mass of an object scales with its size according to $M \approx R^{D^*}$ then for the scattering cross-section from this object $d\sigma/d\Omega \approx Q^{-D^*}$. Thus, a power law regime in Q with an exponent $D^* \leq 3$ relates to a fractal object of dimension D^*. In this sense, an exponent $p = D^* = 1$ relates to rod-like structures, $p = D^* = 2$ to platelets, and $p = D^* = 3$ to open porous objects like networks or house-of-card structures.

Beaucage and Schäfer [47] proposed a general form for the cross-section of a fractal object combining both the Guinier regime where $d\sigma/d\Omega \approx$

$\exp\left(-1/3Q^2R_g^2\right)$ and the fractal regime in a smooth way. For a swollen polymer chain it reads:

$$\frac{d\Sigma}{d\Omega} = G\exp\left(-Q^2R_g^2/3\right) + B\left\{\left[\mathrm{erf}\left(WQR_g/\sqrt{6}\right)\right]^3/Q\right\}^p, \tag{20}$$

with a $Q^{-1/\nu}$ asymptote, a cross-over length $\sqrt{6}R_g$ equal to the end-to-end distance of the chain and a W value of 1.06. The amplitudes G and B are related to each other $B = G\nu/R_g^{1/\nu}\Gamma(\nu)$, where Γ is the Γ function. G is the classical Guinier prefactor:

$$\begin{aligned} &B = G\nu/R_g^{1/V}\Gamma(\nu)\,, \\ &G = \phi(1-\phi)V_p(\rho_p-\rho_s)^2\,, \end{aligned} \tag{21}$$

where V_w is the chain molar volume.

Aggregates representing large three-dimensional objects with sharp interfaces give rise to Porod scattering [48]:

$$\frac{d\Sigma}{d\Omega} = \phi_{agg}2\pi\Delta\rho^2\frac{S_V}{V}Q^{-4} = P_4Q^{-4}\,, \tag{22}$$

Here S_ν is the surface, V the volume, and Φ_{agg} the volume fraction of the aggregates, while P_4 represents the so-called Porod constant. Exponents $3 < p < 4$ in Eq. 19 finally are related to surface fractals [49], while $p > 4$ is characteristic for diffuse interfaces [50, 51].

2.4
Platelet-Like Aggregates with Internal Structure

The novel commercially used wax crystal modifier is based on diblock copolymers with a crystallizable and an amorphous component. These diblock copolymers form platelet-like aggregates in oil solution. Therefore, here we will present in some detail the scattering from such hairy platelets in order to make the later reasoning more transparent. It is easy to write down the scattering from a homogeneous rectangular tile with the dimensions L_x, L_y, and d:

$$\begin{aligned} \frac{d\sigma}{d\Omega}(\underline{Q}) &= \Delta\rho^2\left|\int_0^d\int_0^{L_y}\int_0^{L_x} e^{-\underline{Qr}}\,d^3r\right|^2 \\ &= \Delta\rho^2R(d,Q_z)R(L_y,Q_y)R(L_x,Q_x)\,, \end{aligned} \tag{23}$$

with the function R:

$$R(a,Q) = 4\left[\sin\left(\frac{a}{2}Q\right)/Q^2\right] \tag{24}$$
$$\approx a^2\left(\frac{2\pi}{\sigma}\right)^{1/3}\exp\left[-\left(\frac{3}{\pi}\right)^{1/3}\frac{a^2Q^2}{12}\right].$$

This function has narrow distributions in reciprocal space if the real space extension is large. For large platelet dimensions L_x and L_y the Gaussian approximation for R applies. This is reinforced for platelets with shapes that are neither uniform nor rectangular such that the oscillations of the sin function are averaged out. Furthermore, the relevant intensity seen in the isotropically averaged scattering stems from the region $|LQ| < \pi$.

The reciprocal space intensity is displayed in Fig. 7. It has the shape of a rod along the Q_z axis. Any modulation of the platelet density profile along the z axis yields a form factor:

$$P_z(Q_z) = R(d,Q_z)^2 = \left|\int \phi(z)e^{-izQ_z}\,dQ_z\right|^2, \tag{25}$$

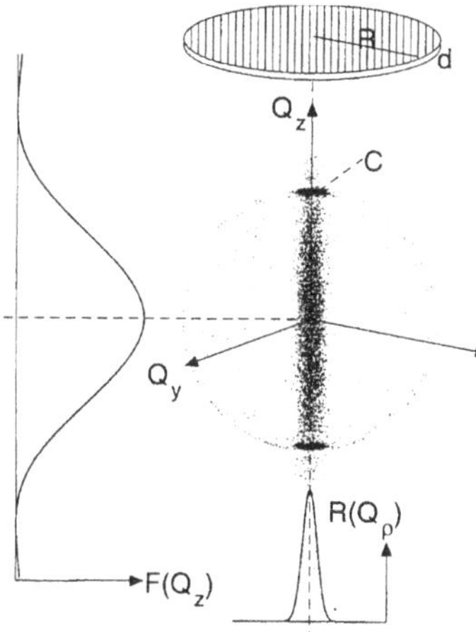

Fig. 7 Reciprocal space intensity from a platelet of finite thickness and large but not infinite lateral extension. The intensity is located on a rod along Q_z, which is modulated by the form factor $|R(Q_z)^2|$ of the density profile perpendicular to the surface. The thickness of the rod is indicated by the Gaussian approximation $R(Q)$. Its width is inversely proportional to the lateral extension. The sphere represents the locations at which the reciprocal space intensity matches the instrumental acceptance if set to a Q value equal to the radius of the sphere. Each dot on the sphere may be considered as one orientation present in a random ensemble of platelets, i.e., an isotropic solution. The overlap of the sphere and the rod, indicated by C, yields the measured intensity. Due to increase of the sphere surface $\sim Q^2$, the intensity combination due to C is $\sim Q^{-2}$ as long as C is only a small spot on the sphere, as in this figure

where $\Phi(z)$ is the lateral platelet profile. $P_z(Q_z)$ modulates the intensity of the rod along its axis for a finite width of the platelet. The thickness of the rod is determined by L^{-1} the lateral extension of the plate.

For an isotropic ensemble of platelets, the detection of scattering takes place on a sphere of radius Q in the platelet-associated coordinate system shown in Fig. 7 [7]. Only if the Q sphere intersects the rod, will intensity contributions be expected. Thus, the detection element is diluted on the Q surface by a factor of $(4\pi Q^2)^{-1}$. For small enough Q, this immediately yields the Q^{-2} behavior of the scattering behavior from thin plates as invoked from the fractal considerations above. Such a behavior implies a divergence at Q equal zero. For finite plates this is avoided, however, by the total immersion of the very small Q sphere within the rod. There the finite inner intensity contribution of the rod is accumulated completely for all orientations. This then yields a scattering intensity proportional to the size of the aggregates. This situation is depicted in Fig. 8.

The disappearance of the Q^{-2} divergence may also be seen if we consider a finite disc with a lateral extension R much larger than the sheet thickness. The orientationally averaged intensity from such a disc is given by:

$$I(Q) = \int_0^{\pi/2} \left| \frac{2\pi R^2 J_1(QR\sin\vartheta)}{QR\sin\vartheta} \right|^2 \sin\vartheta \, d\vartheta \,, \tag{26}$$

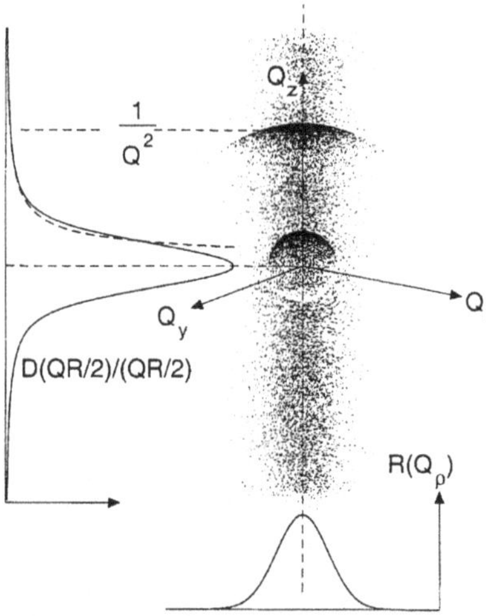

Fig. 8 Enlarged view on the intensity rod in reciprocal space as shown in Fig. 7. Here the overlap function $D(QR/2)/(QR/2)$, *solid line* on the left (see text) and its limiting Q^{-2} behavior are plotted. The modulating form factor $|F(Q_z)^2|$ is virtually constant in this regime. The situation for a larger Q-sphere in the Q^{-2} regime and that for a largely immersed sphere is depicted and may be associated with the corresponding parts of the $D(QR/2)/(QR/2)$ function

with J_1 the first order cylindrical Bessel function. Using a Gaussian approximation for the integrant, Eq. 26 yields:

$$I(Q) = (\pi R^2)^2 \frac{D(QR/2)}{QR/2}, \tag{27}$$

with the Dawson function:

$$D(u) = \exp(-u^2) \int_0^u \exp(t^2)\,\mathrm{d}t \tag{28}$$

$D(u)$ exhibits the following asymptotic behavior. For $u \to \infty$, $2D(u)/u \to 1/u^2$ and for $u \to 0$, $D(u)/u = 1$. Equation 27 interpolates between a constant finite forward scattering and the asymptotic high Q behavior proportional to $1/Q^2$. Thus, a proper consideration of the finite size removes the divergence of the scattering intensity for $Q \to 0$. Necessary modifications for elongated platelets are discussed in [7].

As we shall see, the hairy platelets formed by the crystalline-amorphous diblock copolymers consist of an inner crystalline core and an outer brush. This lateral profile gives rise to a form factor that modulates the profiles of the Bragg rods. The form factor relates directly to the volume fractions $\Phi_\mathrm{B}(z)$ of the brush and $\Phi_\mathrm{C}(z)$ of the core, where z is the coordinate perpendicular to the platelet surface. In terms of these quantities the core brush form factor is given by:

$$\begin{aligned} P(Q) &= \left| \int_{-\infty}^{+\infty} \left[\Delta\rho_\mathrm{b}\Phi_\mathrm{B}(z) + \Delta\rho_\mathrm{c}\Phi_\mathrm{C}(z)\right] \mathrm{e}^{\mathrm{i}Qz}\,\mathrm{d}z \right|^2 \\ &= \Delta\rho_\mathrm{b}^2 P_\mathrm{bb}(Q) + 2\Delta\rho_\mathrm{b}\Delta\rho_\mathrm{c}P_\mathrm{bc}(Q) + \Delta\rho_\mathrm{c}^2 P_\mathrm{cc}(Q)\,, \end{aligned} \tag{29}$$

$\Delta\rho_\mathrm{b}$ and $\Delta\rho_\mathrm{c}$ are the scattering length contrasts between the core and the brush and the solvent, respectively. P_bb, P_bc, and P_cc are the partial form factors for the brush, the interference between brush and core, and the core, respectively. For an explicit modeling of the scattering data, both core and brush were approximated by rectangular shapes convoluted by Gaussians. This has the effect of rounding edges and, if the widths of the Gaussian become large, a continuous transition from a rectangular brush (Alexander and de Gennes [52, 53]) to a more parabolic-like profile is provided [54]. Such a parabolic profile is expected from self-consistent field calculations. Fourier transformation of such profiles yields:

$$\phi_c(Q) = \frac{d \sin(Qd/2)}{(Qd/2)} \exp\left(-\frac{1}{2}Q^2\sigma_c^2\right) \tag{30}$$

$$\phi_b(Q) = m_s \frac{d}{L_b}\left(2(L_b+2)\frac{\sin\left(Q(L_b+d/2)\right)}{Q(L_b+d/2)} \exp\left(-\frac{1}{2}Q^2\sigma_b^2\right) - d\frac{\sin(Qd/2)}{(Qd/2)} \exp\left(-\frac{1}{2}Q^2\sigma_c^2\right)\right).$$

Here σ_c and σ_b are the Gaussian widths describing the core and brush smearing, respectively. d is the core thickness and L_b the brush length. The factor $m_s d/L$ describes the amorphous polymer fraction in the brush. m_s is the geometrical factor describing the fraction of amorphous polymer in the brush. Finally, combining Eqs. 27, 29, and 30 the macroscopic cross-section

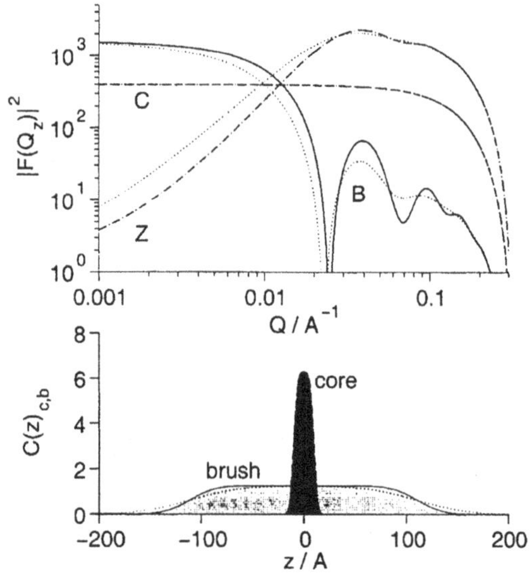

Fig. 9 Form factor contribution $P(Q)$ of the plate with brushes on both sides for different contrasts. C is core contrast ($\Delta\rho_b = 0$), B brush contrast ($\Delta\rho_c = 0$), and Z approximate zero average contrast, ($[\phi_b(Z)\Delta\rho_b + c_c(Z)\Delta\rho_c]\ d_z \cong 0$). The *lower part* of this figure displays the projected average volume fractions of brush $c_b(z)$ and the core $c_c(z)$. The *dotted lines* indicate the results for a larger Gaussian smearing of the brush. The form factor in core contrast stays nearly constant below $Q = 0.1$ Å^{-1}, whereas the brush contrast factor has a sharp minimum around $Q = 0.025$ Å^{-1}, which is directly related to the brush extension

for isotropically oriented platelets becomes:

$$\frac{\mathrm{d}\Sigma}{\mathrm{d}\Omega} = \phi \frac{\phi_{\text{cryst}}}{V_{\text{cryst}}} P(Q)(\pi R^2)^2 \frac{D(QR/2)}{QR/2}, \tag{31}$$

where ϕ is the total polymer volume fraction, ϕ_{cryst} the volume fraction of the crystalline diblock, and $V_{\text{cryst}} = \pi R^2 d$ is the scattering volume of the platelet core.

Figure 9 displays schematically the partial scattering function P_{rs} in Eq. 29 for different contrast situations. While the core form factor extends to high Q and has a simple shape, due to the presence of brushes on both sides of the core, the brush form factor displays pronounced oscillating behavior with the first minimum relating to the average distance between the two brushes. Finally, the dashed dotted line (Z) displays the form factor at approximate average contrast zero. At the lower part of Fig. 9 the perpendicular platelet density profile is displayed.

2.5 Rod-Like Structures with Longitudinal Density Modulation

Random crystalline-amorphous copolymers have the tendency to aggregate in rod-like structures exhibiting a density modulation along the rod (see Sect. 5.1). The form factor appropriate for an ensemble of isotropically oriented cylinders is given by the orientational average of $A(Q, R, L, \alpha)$, the Fourier transform of a cylinder with α as the angle between its axis of symmetry and the scattering vector Q [55]:

$$F(Q, R, L) = \int_0^{\pi/2} A(Q, R, L, \alpha)^2 \sin\alpha \, \mathrm{d}\alpha, \quad \text{with} \tag{32}$$

$$A(Q, R, L, \alpha) = \frac{1}{V} \int_0^R \int_{-L/2}^{L/2} \exp(\mathrm{i}[Qr\sin\alpha + \ell Q\cos\alpha]) r \, \mathrm{d}r \, \mathrm{d}\ell .$$

The integral may be decomposed into a product of radial and axial terms:

$$A_{cs}(Q, R, \alpha) = \frac{2}{R^2} \int_0^R \exp(\mathrm{i}Qr\sin\alpha) r \, \mathrm{d}r, \tag{33}$$

$$A_L(Q, L, \alpha) = \frac{1}{L} \int_{-L/2}^{L/2} \exp(\mathrm{i}Q\ell\cos\alpha) \, \mathrm{d}\ell .$$

The case of a density modulation (alternating more or less dense region) along the cylinder direction can be described by modifying the axial term as follows [26]:

$$A_L(Q,L,h,\alpha) = \frac{1}{L}\int\limits_{-Nh}^{+Nh} \exp(\mathrm{i}Q\ell\cos\alpha)\left\{2(1-c)\left[\cos\left(\frac{2\pi}{h}\ell\right)\right]^2 + c\right\}\mathrm{d}\ell\,. \tag{34}$$

The *cosin* function describes the modulation with a period h over the cylinder of length $L = 2Nh$. c is the average amplitude of the modulation. Unfortunately Eq. 32 and the following cannot not be solved analytically. Nevertheless, a numerical solution can describe the salient features of the neutron cross-section, from an ensemble of cylinders of radius a and length L, which are randomly oriented and display a modulated density along the axis.

2.6 Scattering from Polymer Brushes – Blob Scattering

So far, we have considered homogeneous segment density distributions in the brush structures. However, as in semidilute polymer solutions in good solvents, the screening of the excluded volume interaction is expected to produce blob structures. They are also at the basis of the theoretical understanding of polymer brushes due to Alexander and de Gennes [52, 53].

Blob scattering results from fluctuations of the polymer segment density in the brush. It relates, thus, to disorder and prevents interferences with the scattering amplitudes from the average core and brush densities. This disorder scattering can be obtained directly from the excess scattering beyond that expected from homogeneous density distributions. Independent of any model a core brush structure with homogeneous density distributions requires (in analogy to Eq. 11):

$$S_{\mathrm{bc}}^2(Q) = S_{\mathrm{bb}}(Q)S_{\mathrm{cc}}(Q)\,, \tag{35}$$

where the subscripts b and c stand for brush and core respectively. With that the excess brush scattering $I(Q)_{\mathrm{exc}}$ can be formulated as [56]:

$$I(Q)_{\mathrm{exc}} = \phi\Delta\rho_{\mathrm{b}}^2\left[S_{\mathrm{bb}}(Q) - \frac{S_{\mathrm{bc}}(Q)^2}{S_{\mathrm{cc}}(Q)}\right] = \phi\Delta\rho_{\mathrm{b}}^2\tilde{S}_{\mathrm{bb}}(Q)\,. \tag{36}$$

Therefore S_{bb}, S_{bc}, and S_{cc} are the experimentally obtained partial scattering functions and $\tilde{S}_{\mathrm{bb}}(Q)$ is the partial scattering function for fluctuation or blob scattering. Equation 36 assumes an experiment under brush contrast.

In the simplest case of blob scattering from an ensemble of identical blobs, the correlation function for one blob may be described as:

$$g(r) \approx r^{1/\nu-3}\exp\left(-\frac{r}{\xi}\right)\,, \tag{37}$$

with ν the Flory exponent and ξ the mean blob size. Fourier transformation and normalization results in:

$$I(Q)_{\text{exc}} = \phi\alpha\Delta\rho_{\text{b}}^{2}\frac{\sin(\mu\arctan(Q\xi))}{Q\xi(1+(Q\xi)^{2})^{\mu/2}}\,. \tag{38}$$

with the normalization constant $\alpha = V_{\text{brush}}m_{\text{s}}(d/L)\xi^{3}(1/\mu)$. There, m_{s} is the relative fraction of the amorphous polymer in the diblock copolymer and $\mu = 1/\nu - 1$. Finally, according to Alexander and de Gennes [52, 53] the brush size is given by the mean distance between the brush-forming polymers at the surface of the aggregate. With that we obtain:

$$\xi = \left(\frac{2M_{\text{cryst}}}{\zeta_{\text{cryst}}N_{A}d}\right)^{2}, \tag{39}$$

where N_A is the Avogadro number, M_{cryst} the molecular weight of the crystalline part of the block copolymer and ζ_{cryst} the density of the crystalline aggregate.

2.7 Structure Factors

As long as the solutions are sufficiently dilute such that the aggregates are well separated one expects little in the way of correlations of the aggregate center orientations. Under these circumstances, Eq. 31 or analogs of it are complete, since the scattering from different platelets takes place without interference effects. However, if the concentration is increased, correlations become more and more important yielding an extra factor, $S(Q)$, the structure factor. For non-spherical particles, $S(Q)$ is very complicated and intractable for all purposes. An exception is the stacking of large platelets, which for example could be driven by the van der Waals interactions between large flat surfaces. The arising structure factor has to be multiplied by the single platelet scattering function. If we would have a perfect stack, then the intensity that is distributed on rods perpendicular to the platelet surfaces would condense into points on the rods. The distance between which is $RV = \Delta Q_{\text{normal}} = 2\pi/D$, where D is the stacking period in real space and RV would be the corresponding reciprocal lattice vector. For randomly oriented staples the scattering pattern would be a sequence of Debye–Scherrer rings. We now consider a random variation in the distance of neighboring platelets and treat the arising structure factors in terms of a one-dimensional paracrystalline structure. This assumes a Gaussian probability distribution of the next neighbor distance, leading to a distribution function:

$$w_n(z, D) = \frac{1}{\sqrt{\pi n\sigma_D^2}}\exp\left(-\frac{1}{N}\frac{(z-nD)^2}{\sigma_D^2}\right), \tag{40}$$

for the n-th neighbor distance, because of the convolution properties of Gaussian distributions. Fourier transformation of the one-dimensional Patterson function $\sum w_n(D)$ leads to:

$$S(Q) = \sum_{n=-\infty}^{\infty} e^{iQDn - nQ^2\sigma_D^2/4} = \frac{\sin h\left(Q^2\sigma_D^2/4\right)}{\cos h\left(Q^2\sigma_D^2/4\right) - \cos(QD)} . \tag{41}$$

Here $S(Q)$ is the structure factor of the one-dimensional paracrystalline order [57]. It is easily verified that $S(Q \to \infty) = 1$, as required for the structure factor of any disordered systems. The effect of the distance variation is a dephasing of the interference terms, reducing the effect after ensemble averaging. For high enough Q the interference is completely destroyed. The same effect may occur by a random mutual tilt of different platelets in a stack resulting in a tilt of the corresponding Bragg rods. The scattering from any two platelets can only interfere if their Bragg rods overlap. We may easily imagine that for thin Bragg rods this overlap goes to zero with increasing Q as soon as they are not perfectly parallel. The effect of this additional interference loss is not easily discernable from the dephasing due to distance variations. Therefore, we include such effects in an effective value of σ_D, which may come out larger than expected from pure distance variations.

Up to now we have implied a homogeneous filling of space by a stack, or at least very large stacks. For finite stacks we have to further modify $S(Q)$. In this case Eq. 41 changes to:

$$S_N(Q) = 1 + \frac{2}{N} \sum_{n=1}^{N_p} (N_p - n) \cos(QDn) e^{-Q^2\sigma_D^2 n/4} , \tag{42}$$

where N_p is a number of platelets in the stack. Figure 10 illustrates the results.

Figure 10a displays schematically a stack of platelets. Figure 10b compares the structure factors $S_N(Q)$ and $S(Q)$. The main effect of the structure factor is a strong reduction of the intensity for small Q ($Q < 2\pi/D$). Finite stacks lead to strongly increasing intensities below $Q < 2\pi/(ND)$; this scattering may be understood as the forward scattering from the whole aggregate.

3
Ethylene/Vinylacetate (EVA) Copolymers

In this section we discuss some recent combined SANS and yield stress studies of EVA fuel additives that have a widespread use in middle distillate fuels [58, 59]. Other than the later discussed diblock copolymers or random crystalline-amorphous copolymers, relatively little has been done with respect to the microscopic evaluation of the polymeric aggregates nor on their interaction with wax by scattering techniques. These results nevertheless are presented first, in order to set the scene and to have a reference for the later

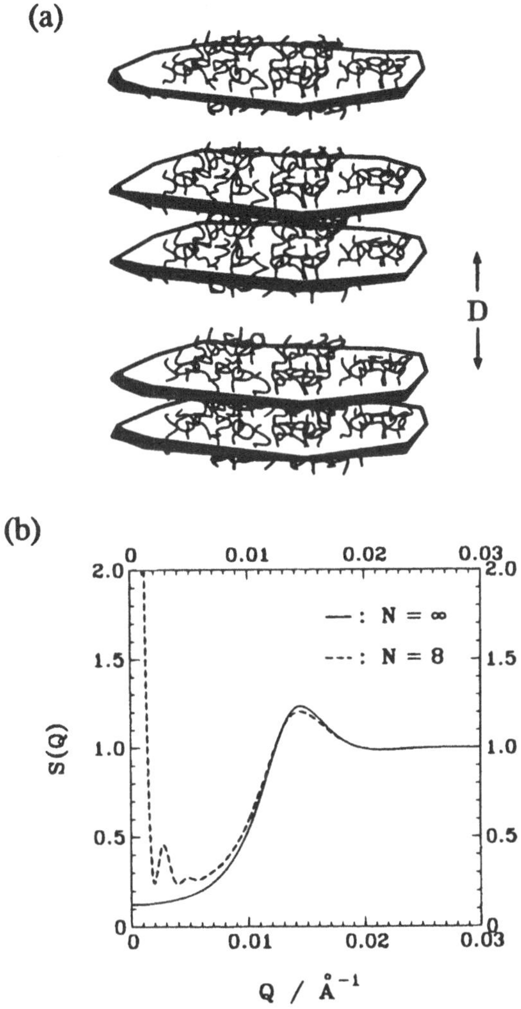

Fig. 10 **a** Schematic sketch of an aggregate stack. **b** Resulting structure factors $S_N(Q)$ for a finite and $S(Q)$ for an infinite staple calculated for an average interlamellar distance of $D = 400$ Å and a Gaussian smearing of $\sigma_D = 200$ Å

discussion of the novel wax crystal modifiers, which are the main subject of this review.

The ethylene vinylacetate copolymers are chains with polyethylene sections of varying length depending on the amount of copolymerized vinylacetate (VA) comonomer. This family of materials has found widespread use as a pour point depressant additive for crude oils and petrochemical products, with varying degrees of success in controlling the size and morphology of the wax crystals formed. When successful as a wax crystal additive, EVA produces crystals that are much smaller and more numerous than those found in the unmodified system [2, 60]; the crystals are also more compact [61].

In spite of many investigations carried out by viscosity measurements [62, 63], DSC [58, 60, 64], reaction calorimetry [2], infrared spectroscopy [64], X-ray diffraction [64], and computer simulation [65, 66] the mechanism of action of the EVA copolymer is still not resolved. The two mechanisms of heterogeneous nucleation and growth inhibition were proposed, but without convincing evidence in favor of one or the other [67–69]. EVA copolymer as the nucleating agent should crystallize from the solution above the crystallization temperature of the waxes. However, measurements of cloud points gave, sometimes, contradictory results revealing either a depression [59] or an increase of the CP [60]. Also, a growth inhibition mechanism has been put forward as the result of several experimental investigations and simulations of polymer flow improvers with long alkyl lateral chains [70–72]. Moreover, it seems that EVA copolymers can act either as nucleating agents or as crystal inhibitors depending on the VA content of the copolymer, the composition of the waxes, and the chemical nature of the solvent. Thereby, the efficiency of the EVA copolymers is not always the same. Their performance is generally limited to a certain composition of oils and fuels [59]. On the other hand there is experimental evidence that the molecular parameters of the EVA copolymer play an important role in the performance and the action mechanism of these copolymers as wax crystal modifiers [73, 74]. For example, it seems that the efficacy of EVA as a wax crystal modifier is increased by increasing the extent of intrachain to interchain interaction [75]. This balance of associations is affected by the polymer molecular weight and degree of VA content as well as the thermal history of the polymer solution. However, it is difficult to correlate the aggregation properties of EVA and its wax modification capacity because of the poorly characterized nature of these polymers due to the pronounced distribution in molecular weight and composition along with the potential for chain branching [76].

A better understanding of the aggregation behavior of EVA copolymers acting as wax modifier additives was recently achieved by studies combining structural and rheological investigations as a function of temperature, wax composition, and VA content [10, 18]. The work was performed on the different commercial grades of EVA additives where the VA content varies between 9% and 15%.

The temperature evolution of the self-assembly of all three EVA grades and the aggregated structures in decane were investigated by SANS at the SANS instrument KWS-1 at the FRJ-2 research reactor in Jülich/Germany. The range of momentum transfer varied between $0.002\ \text{Å}^{-1} < Q < 0.2\ \text{Å}^{-1}$.

The scattering patterns from all EVA samples in 2% decane solution are shown in Figs. 11 to 13. We note the increase of the scattering with decreasing temperature, whereby the observed temperature dependencies as well as the shapes of the scattering patterns are significantly different for the different

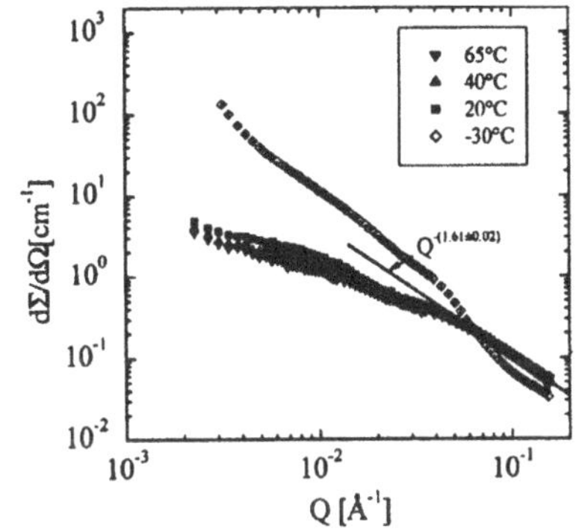

Fig. 11 Scattering results of EVA-ga 2% solution at different temperatures

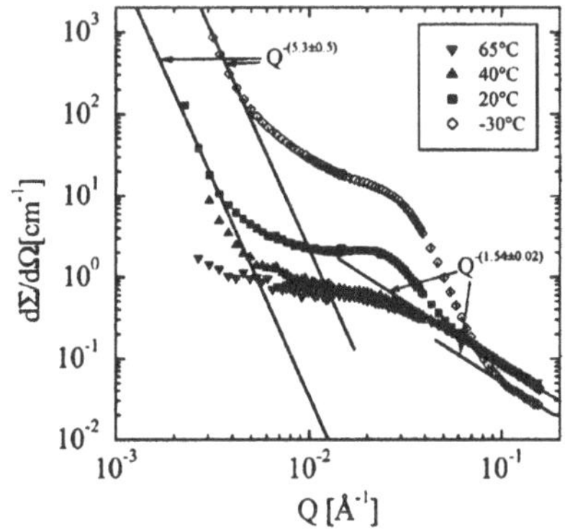

Fig. 12 Scattering results of nucleators-grade ethylene/vinylacetate (EVA-nu) 2% solution at different temperatures

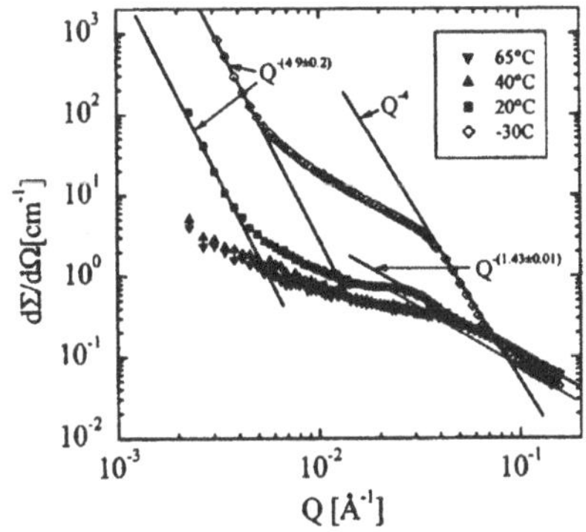

Fig. 13 Scattering results of EVA-ss 2% solution at different temperatures

materials. The nucleator grade shows the strongest scattering and the growth arrestor shows the weakest scattering, indicating significantly different aggregation properties of the three polymers. As the common feature we note a significant decrease of the scattering intensity at high momentum transfers ($Q > 0.07$ Å^{-1}) indicating a significant fraction of "inactive" material.

The scattering patterns from the EVA-ga (the polymer with the highest VA content) are displayed in Fig. 11. The intensity profiles for room temperature and above are almost identical. They display a $Q^{-5/3}$ tail at high Q indicating the scattering from a swollen coil. At −30 °C a strong intensity increase is observed that exhibits a power law with an exponent of ~1.6 decaying to that of a very open fractal aggregate of a dimension of $D = 1.6$ (see Eq. 19). The above-mentioned intensity decrease at high Q most likely relates to the fact that a fraction of the EVA molecules precipitates out of solution prior to any wax interaction. Indeed, at room temperature the visually inspected solution was opaque and a white sediment accumulated at the bottom of cell.

The EVA-nu contains a lower VA content and thus longer methylene unit sections with a higher potential for crystallization and therefore aggregation. This behavior is clearly revealed by the scattering pattern of Fig. 12. A distinct increase in the scattering intensity already begins at temperatures around 40 °C. Thereby the strong increase of scattering at low Q is particularly noteworthy. Since only a narrow Q range in the low Q regime is covered, a determination of the power law is quite difficult. Nevertheless, exponents of $\rho > 5$ are seemingly present. Such strong power laws cannot be interpreted in terms of fractal aggregates or fractal surfaces and most likely originates from diffuse surface scattering where the Porod exponent (see Eq. 22) is further increased by the smeared out surface. We also note that even for the smallest Q values no saturation of the power law behavior is visible. With that we may conclude that EVA-nu forms quite large aggregates with pronounced surfaces and aggregate sizes well above several thousand Angstroms. This pronounced surface scattering distinguishes the EVA-nu aggregates from the EVA-ga where such behavior is not visible. Thus, EVA-nu appears to develop much more compact structures than EVA-ga implying that the surface area available for wax nucleation is clearly diminished.

Another feature in the scattering pattern is the presence of an intermediate peak at $Q^* \cong 0.025$ Å^{-1}. Such a peak indicates a structural correlation length of $\xi = 2\pi/Q \cong 250$ Å of a still unclear microscopic identification. Finally, the loss of intensity at high Q and low temperature is by far the most pronounced of the three grades. It most likely originates from both precipitation as well as from the fact that part of the material is embedded in a more compact structure giving rise to a loss of polymer scattering.

The EVA-ss copolymer (Fig. 13) contains a VA fraction that is intermediate between those of the two other EVA samples. Therefore, an intermediate aggregation and scattering behavior is expected. While at 65 °C the scattering pattern of the EVA-nu and the EVA-ss are very similar, at 40 °C the EVA-nu already displays signs of significant aggregation whereas the ss grade does not. Here significant aggregation begins between 40 °C and 20 °C. At room temperature and low Q a very strong intensity increase is observed ($\rho \cong 4.9$), which again is indicative of a fraction of large particles with diffuse interfaces. At –30 °C the ss and the nu grade polymers again exhibit rather similar scattering patterns with a generally lower intensity level of the ss grade. As for the EVA-nu grade, again there are signs of an intermediate scale of ~250 Å.

In summary, the EVA-nu and EVA-ss materials display distinctly different aggregation behavior to EVA-ga. While the latter only weakly aggregates and forms very open fractal structures, the other two materials aggregate into large scale compact structures with a defined but diffuse surface. Although these large aggregates make the solutions totally opaque to white, the transmission of neutrons is only slightly changed from 0.78 at room temperature to 0.73 at –30 °C. In addition, as already mentioned, as room temperature is approached a white sediment slowly accumulates at the bottom of the cell. This effect correlates with the decrease of the scattering intensity at larger Q, which amounts to 23% in the EVA-ga sample, 38% in the EVA-ss sample, and 55% in the EVA-nu sample. This segregation phenomena at lower temperatures is detrimental to the wax crystal modification activity because the self-assembled structures are removed from the solution.

Finally, we may try an assignment of the correlation peak at intermediate Q that is present both in the EVA-ss and EVA-nu scattering pattern. This peak might relate to the long period Bragg reflection of the orthorhombic crystals formed by the crystalline sections of the EVA materials. Low angle X-ray diffraction performed on EVA containing 12% VA units, which was crystallized by evaporation from a solution in CCl_4, revealed a Bragg peak at $Q = 0.033$ Å^{-1} that was assigned also to the long period of the lamellar structure of semicrystalline EVA [64]. On the other hand, the SANS results on the EVA solutions did not reveal any evidence for scattering from such two-dimensional structures (no sign of an indicative Q^{-2} power law).

The EVA-ga and EVA-ss copolymers have also been investigated with regard to their effect on the yield stress of paraffin/decane solutions that readily form solid-like gels below the wax precipitation temperature. These waxy oil solutions for the measurement of the yield stress were prepared as mixtures of 4 wt % linear paraffin waxes of carbon numbers C_n ($n = 28$, 32, and 36) with variable concentrations of EVA-ss and EVA-ga. EVA-nu was not used in the rheological study because ~55% of the material precipitates from solution according to the SANS observations. The rheological study concerned gel formation under conditions such as those that occur in a diesel engine or a pipeline during shutdown.

Figures 14 and 15 present the effect of EVA-ga and EVA-ss on the relative yield stress of various 4% wax gels in decane at 0 °C. Below the wax CP, paraffin waxes precipitate in platelet-like structures, which at sufficiently high wax concentration form a sample spanning a house-of-cards structure that entrains decane and forms a gel with a large yield stress. EVA breaks down the gel structures of all three paraffins considered. Additive concentrations

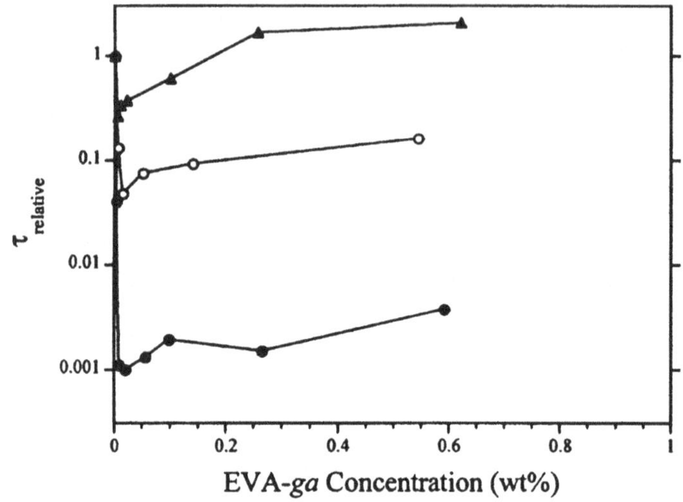

Fig. 14 Effect of growth-arrestor-grade ethylene/vinylacetate (EVA-ga) on the yield stress of 4 wt % paraffin wax gels in decane at 0 °C, as a function of the wax modifier concentration. The relative yield stress is defined as the yield stress of the sample with modifier divided by the yield stress of the sample without modifier. *Symbols* correspond to results for C_{28} (*closed triangles*), C_{32} (*open circles*), and C_{36} (*closed circles*) waxes

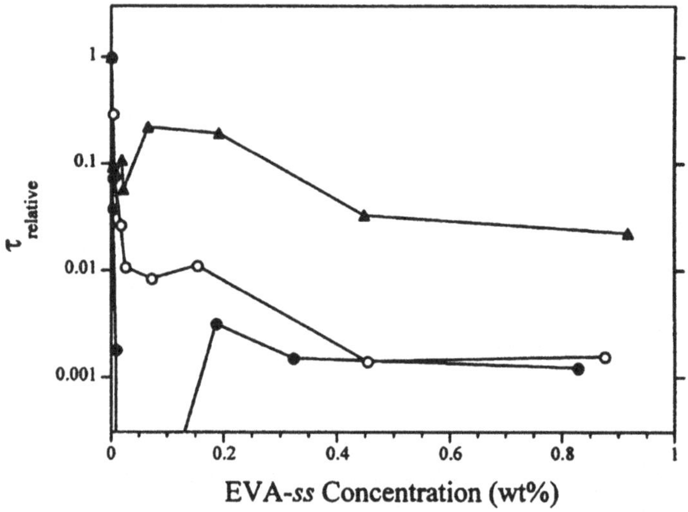

Fig. 15 Effect of single-shot-grade ethylene/vinylacetate (EVA-ss) on the yield stress of 4 wt % paraffin wax gels in decane at 0 °C, as a function of wax modifier concentration; the relative yield stress and *symbols* are as defined in the caption for Fig. 14

of ~100 ppm are sufficient to reduce the yield stress of the C_{36} gels by three orders of magnitude or more.

Both additives display an increasing selectivity for the ability to break down gels formed by paraffins of increasing chain length, i.e., the minimum in the relative yield stress obtained for the C_{36} wax is lower than that observed for the C_{32} wax, which in turn is lower than that observed for the C_{28} wax. Indeed with increasing EVA-ga concentration the yield stress of the C_{28} wax eventually becomes even larger than that observed for the neat wax gel, i.e., the gel is reinforced by the increasing additive concentration. The two wax modifiers display a different ability to break down the wax gels. The selectivity of the EVA-ga is very pronounced and there is a good separation between the yield stress curves for the three waxes considered. In contrast, the EVA-ss grade is the most active for concentrations of about 0.1 wt %; here a minimum in the relative yield stress for all three paraffins is observed. For the C_{36} wax, the yield stress decreases to such low values that in the region of the minimum the rheometer was unable to identify a yield stress. At intermediate concentrations of 0.2 wt % EVA-ss the yield stresses of the various paraffins pass through a maximum before decreasing again to a plateau at higher EVA-ss concentrations.

Surprisingly, a similar selectivity of the EVA-ga and EVA-ss grades for the longest paraffin considered was found. Moreover at high additive concentrations, EVA-ss breaks down both C_{32} and C_{36} gels to a comparable extent whereas EVA-ga displays a greater affinity for breaking down the gel structures of C_{36} compared to C_{32} over the entire range of concentration.

The present SANS and rheological findings revealed that EVA random copolymers most likely promote the nucleation of wax via crystallizable methylene runs and thus reduce the dimension of the wax platelets and collapse the wax house-of-cards. Furthermore, it is clear that the commercial designations of growth arresters, single shot, and nucleators are heuristic and do not explain mechanistically the role of these polymers in modifying the wax crystal structure.

In an Edisonian sense it is known, via CFPP tests, that the EVA-nu and EVA-ss grades can decrease the value of this parameter, thus leading to low temperature serviceability. However, the presumed "growth arresting" capacity of the EVA-ga grade is apparently based upon supposition rather than direct experimental evidence. The EVA-ga is partnered with the EVA-nu in the commercial formulations. The growth arrestor grade could, of course, act as a nucleator in a middle distillate fuel containing, one may surmise, lower chain length waxes. No evidence is available that attests to the presence of wax crystal growth arresting activity.

The efficiencies of the EVA grades are, as noted, hindered by the precipitation from solution at temperatures *prior* to wax formation. Details regarding the characteristics of the EVA grades remaining in solution (which are the parents of the productive self-assembled structures that ultimately serve as the wax templates) is unavailable in spite of the rather obvious value of such information. It seems that these samples could be readily collected from a controlled fractionation of the various EVA grades as a function of temperature. This exercise would provide samples of both the useful and rogue EVAs in the commercial formulations. The near-uniform composition and near-monodisperse molecular distributions of the anionic based samples do not suffer from this spontaneous waste of additive solubility.

4 Crystalline-Amorphous Diblock Copolymers

While EVA wax crystal modifiers have a rather mixed performance, including rather significant fractions of inactive materials, the novel line of wax crystal modifiers based on relatively low molecular weight crystalline-amorphous diblock copolymers have proven to exhibit excellent activity as CFPP suppressors even in unruly fuels [77]. This included a number of middle distillate fuels that resisted any treatment or formulations based on the EVA additives. In the case of the EVA copolymers, the application evaluations were based on an Edisonian approach. In contrast, the development of the novel diblock copolymer additives was based on a microscopic scientific approach, where quantitative knowledge of the aggregation behavior guided the optimization of the additive activity.

We will first describe the structure of the diblock copolymer additives. Then, from the structural data the aggregate free energy of formation is derived. Microscopic investigations also served in order to directly observe the wax additive interaction showing that the available aggregate surface is the decisive parameter. This aggregate surface is then derived on the basis of the free energy of formation and finally compared with experimental measurements of the CFPP.

4.1 Aggregates of Crystalline-Amorphous Diblock Copolymers

The polyethylene (PE)-like component in the PE-PEP diblock crystallizes in solution with a high enthalpy gain and generates the thermodynamic reason for aggregation. From the lamellar morphology of polyethylene crystals, two-dimensional structures are expected for such diblock copolymers with the crystalline polyethylene core surrounded by a PEP or a PEB-*n* amorphous brush. The detailed morphology of such aggregates should be determined by the balance of entropic forces resulting from the stretching of the brush hairs and the enthalpic contribution from the polyethylene chain folding as well as the remaining contribution from the ethyl side branches derived from vinyl groups in the dominantly 1.4-polybutadiene parent structure.

SANS investigations were performed in order to quantitatively understand the aggregation phenomena occurring in solution and to decipher the influence of the PE-PEP additives on the wax crystals formed at low temperatures in middle distillates. PE-PEP diblocks with varying molecular weights and compositions were evaluated [7, 8, 78] as well as combinations with straight chain paraffin waxes C_n of different carbon numbers in decane solvent [8, 78, 79]. These solutions may be considered as models for diesel fuels.

Detailed studies of the structure and morphology of such polymer systems by SANS techniques require the variation of the scattering properties of the different polymer components by varying the degree of hydrogenation or deuteration, respectively. Furthermore, the wax may be contrasted out or made visible again depending on the degree of deuteration of the different components. Finally, the yielding behavior of waxy gels with and without diblock was investigated as for the EVA materials [16]. The wax crystal morphologies were examined using an optical microscope with a Peltier plate mounted on the microscope stage.

4.2 PE-PEP Self-assembling

Below 35 °C an observation of the polymer solutions by phase contrast microscopy reveals aggregates of needle-like shapes extending to several microns. Their thickness is at the border of the microscope resolution, i.e., in the order of 200–500 nm. As stated above from the crystallization properties of PE for the PE-PEP diblock copolymers we would expect lamellar structures containing crystalline PE cores with PEP brushes on both sides. The microscopy result of thicker aggregates points towards stacking of such platelets into larger aggregates. The scattering behavior of such stacks with internal structure is discussed thoroughly in Sects. 2.4, 2.6, and 2.7. Combining Eqs. 31, 36, and 41 we arrive at:

$$\frac{\mathrm{d}\Sigma}{\mathrm{d}\Omega} = \phi_{\mathrm{cryst}} 2\pi \frac{P(Q)}{Q^2 d} S(Q) + I_{\mathrm{exc}}(Q)\,, \tag{43}$$

where d is the thickness of the crystalline core.

Figure 16 displays the scattering data from a diblock copolymer with a notation 5DH/8HH at a volume fraction of $\phi = 2\%$ and room temperature for five different solvent scattering length densities [7]. This notation means that we deal with a deuterated polybutadiene and hydrogenated polyisoprene parent polymer, which was subsequently hydrogenated. The solvent scattering length densities were varied in equal steps between fully deuterated decane ($\rho_\mathrm{S} = 6.2 \times 10^{10}\ \mathrm{cm}^{-2}$) and fully protonated decane ($\rho_\mathrm{S} =- 0.3 \times 10^{10}\ \mathrm{cm}^{-2}$). From such measurements, using the procedures outlined in Sect. 2.2, the partial scattering functions S_cc, S_bb (where c stands for core and b for brush) may be obtained.

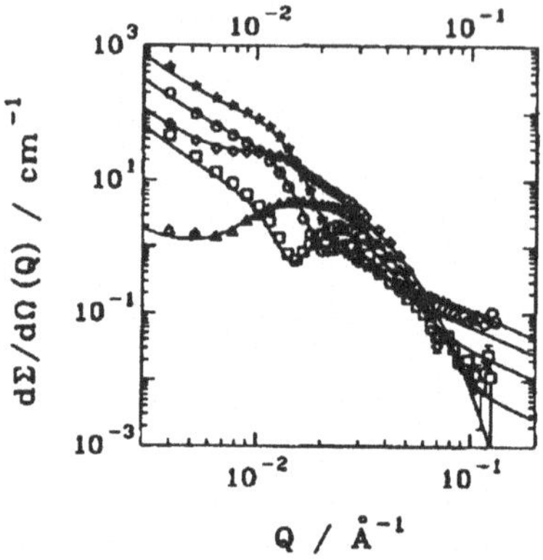

Fig. 16 Data profiles obtained from contrast variation using five different decane contrasts for the diblock 5DH8HH at $\Phi = 2\%$ and room temperature. The *solid lines* display the fit prediction using a smeared step function for the brush profile. The scattering lengths densities of decane corresponding to the different curves are ☆ 6.2×10^{10} cm^{-2}, ○ 4.6×10^{10} cm^{-2}, □ 2.9×10^{10} cm^{-2}, △ 1.3×10^{10} cm^{-2} and ◇ -0.31×10^{10} cm^{-2}. For better visibility only every third data point is displayed

Figure 17 displays such data obtained under core and brush contrast conditions. Such a representation directly yields the product of the form factor and the structure factor:

$$\frac{d\Sigma}{d\Omega}(Q)Q^2 \approx P_{ij}(Q)S(Q)\,, \tag{44}$$

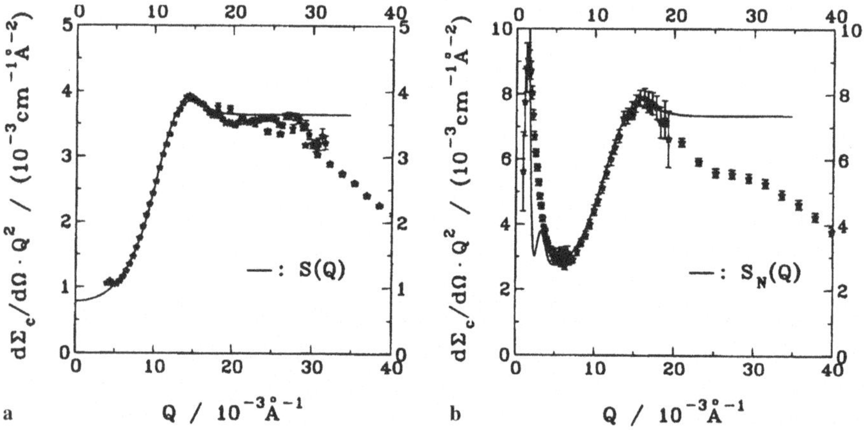

Fig. 17 Kratky plots of the partial scattering function under core contrast from the 5DH8HH (**a**) and 6HH10DH (**b**) samples at $\phi = 2\%$ at room temperature. The *solid lines* are fits with the paracrystalline structure factor for $Q \leq 2.5 \times 10^{-2}$ Å^{-1}. For **b** a finite stapling of $N = 8$ could be determined

which is valid for $Q > 2\pi/R$. We first concentrate on the core contrast conditions. There at low $Q < 1/d$ the form factor of a single platelet is governed by its thickness and is about constant. Then Eq. 44 directly yields the structure factor due to the stacking of platelets. The data in Fig. 17 display a maximum at $Q = 15 \times 10^{-3}$ Å^{-1} indicating an interplatelet distance of $D = 400$ Å. At lower Q the intensity is strongly reduced displaying the typical profile of a structure factor. The solid line in Fig. 17a presents a fit with a paracrystalline structure factor of Eq. 41. From this fit we find an interplatelet distance of $D = 380$ Å, which is accompanied by a large smearing of $\sigma_D = 250$ Å. Figure Fig. 17b displays the partial scattering function for brush contrast for the 6HH/10DH PE-PEP diblock again in a Kratky representation. The data display again a maximum at $Q \cong 15 \times 10^{-3}$ Å^{-1} indicating an interplatelet distance of 400 Å. Below $Q = 5 \times 10^{-3}$ Å^{-1} an additional strong intensity increase is observed. This intensity increase could be due to two different reasons:

(i) Finite platelet dimension: If Q approaches the inverse platelet dimension, R, the Dawson function in a Kratky representation leads to a weak maximum at $QR \cong 3$ with a relative height of about 30%.
(ii) Finite stapling: The observed much stronger increase of the intensity points in the direction of finite aggregates. A fit with Eq. 42 results in microaggregates of about $N_p = 8$ platelets. The corresponding structure factor is included in Fig. 17b. From that a thickness of 8×40 nm $= 320$ nm may be calculated comparing very well with the needle thickness at the border of the resolution in the phase contrast microscope.

The scattering pattern at higher Q is informative in terms of the form factors of both core and brush. For $Q > 0.02$ Å^{-1} the influence of the interplatelet structure factors become small (see Fig. 17); $S(Q)$ essentially approaches one. Following from Eq. 44, the form factor of the core and brush structures are observed. For the core, it follows from Eq. 30 that:

$$P_{cc}(Q) \approx \frac{\sin(Qd/2)}{(Qd/2)^2} \exp\left(-Q^2\sigma_c^2\right) . \tag{45}$$

Figure 18 displays a fit to Eq. 45 of the core contrast data from sample 6HH10DH in the Q range above 0.02 Å^{-1}. Any interplay between structure and form factor is neglected. Such a fit yields a core thickness of $d = 52$ Å for this sample. The insert in Fig. 18 shows the corresponding density profile in real space. We emphasize the relatively strong smearing of the density profile with a Gaussian width of $\sigma_c = 18$ Å. The large value of σ_c traces back to the relatively strong decrease of the scattering profile towards higher Q. The behavior is systematic for all investigated diblock aggregates and indicates a rather rough surface structure.

For the brush, with its larger length scales, the separation of scales between those of the structure factor and that of the brush are not as well fulfilled as in the case of the core. Therefore, its profile cannot be obtained

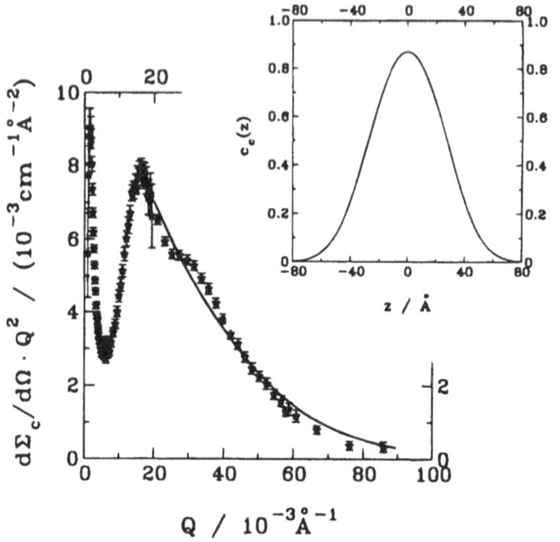

Fig. 18 Fit of the core form factor $P_{cc}(Q)$ to the partial scattering function under core contrast from the diblock 6HH10DH at $\Phi = 2\%$ at room temperature ($Q \geq 25 \times 10^{-3}$ Å^{-1}). The *insert* displays the corresponding density profile of the PE core

in a semi-quantitative way as before but needs an explicit consideration of the structure factor. Furthermore, in the brush beyond the average density profile concentration, fluctuations are also important. This has been discussed in Sect. 2.6 where the excess scattering from the monomer correlations within the brush was discussed. From Eq. 43 in combination with Eq. 30 the macroscopic cross-section for brush scattering in the Kratky representations becomes:

$$\frac{\mathrm{d}\Sigma}{\mathrm{d}\Omega}(Q)Q^2 \approx P_{\mathrm{bb}}(Q)S(Q) + I_{\mathrm{ex}}(Q)Q^2 \,. \tag{46}$$

P_{bb} is the form factor of the homogenous brush profile while I_{ex} is the contribution from the concentration fluctuation in the brush. We note that this diffuse scattering is not subjected to the structure factor. Figure 19 displays experimental partial structure factors for brush contrast in the Kratky format. Compared to the core scattering profiles (Fig. 18) these data are richer in structure and display two different regimes. At low Q, as a result of the interplay between $S(Q)$ and the form factor, a first peak appears around $Q = 1 \times 10^{-2}$ Å^{-1}. This peak is followed by a deep minimum around $Q = 0.02$ Å^{-1}. Interference effects cause this pronounced minimum from the scattering of the two brushes on both sides of the PE plate core. Its position is a measure of the distance S between the center of masses of the two brushes, $S \cong \pi/Q \cong 160$ Å. The fit distinguished between two different types of brush profiles. The Gaussian profile as described in Eq. 30 (full line) and

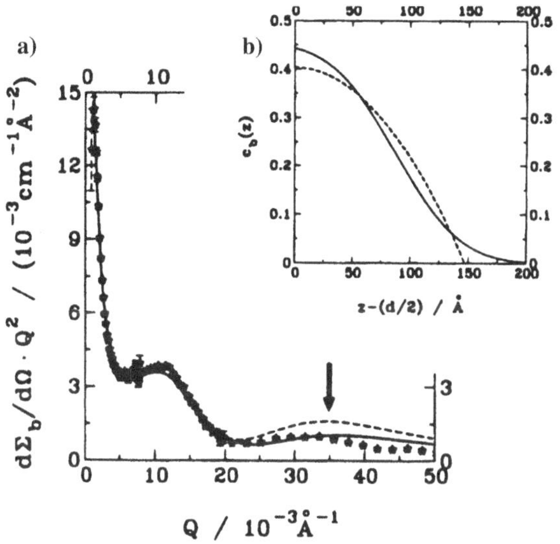

Fig. 19 Kratky representation of the partial scattering function under brush contrast for the diblocks **a** 5DH8HH and **b** 6HH10DH, each at $\Phi = 2\%$ at room temperature. The *solid* and *dashed lines* display fits with $P_{bb}(Q)\, S(Q)$ using different brush profiles. The *inserts* display the corresponding polymer densities in real space (*solid line*, smeared step function; *dashed line*, parabolic profile)

a parabolic profile as predicted from self-consistent field calculations (dashed line). In Fig. 19 both best fits are compared to the experimental results. In both cases the experimental data are well described except for deviations occurring at large Q. While in the case of the 5DH8HH diblock copolymer both profiles lead to an equivalent description for the 6HH 10DH system, the smeared step function profile is favored. The inserts compare the respective density profiles in real space.

Using the full expression for the structure factor Eq. 43 the data profiles from a number of different diblock copolymers have been fitted simultaneously for different scattering contrasts [7]. From such fits a very strict determination of the structural parameters becomes possible. The solid lines in Fig. 16 display the results of such a fit for the core and brush parameters. In the case of the brush we display both the parameters for a smeared step function as well as those for a parabolic profile. Table 1 displays the core and brush parameters, while Table 2 presents the corresponding structure factor results.

Though the form factor parameters of the single lamellar structure and the structure factor parameters have been determined independently, we note that in all cases the relation $2L_p + d \cong D \pm \sigma_D$ holds. Obviously the stacking period D coincides with a lateral aggregates size. We also note that at a first glance the very thin core thickness of $d \cong 50$ Å is quite astonishing – d is con-

Table 1 Core and brush parameters from the combined fits (see text) for brush and core dimensions

Diblock			Smeared function		Parabolic profile[a]
	d	σ_d	L	σ_L	L_P
5DH8HH (2%)	38 ± 3	21 ± 3	123 ± 2	42 ± 2	183 ± 2
5DD10HD (1%)	32 ± 4	16 ± 3	124 ± 3	67 ± 3	190 ± 3
6HH10DH (1%)	52 ± 4	17 ± 2	103 ± 3	48 ± 3	167 ± 2
6HH10DH (2%)	52 ± 4	17 ± 2	93 ± 5	46 ± 4	155 ± 2
10DH16HH (2%)	80 ± 4	6 ± 1	225 ± 2	81 ± 2	334 ± 2
6HH15DH (2%)	44 ± 4	23 ± 3	186 ± 7	96 ± 7	303 ± 3
6DH20HH (2%)	32 ± 2	15 ± 2	231 ± 5	72 ± 5	333 ± 4

For the brush, both the parameters for a smeared step function profile as well as those for a parabolic profile are included [a] L_P should be compared to $L + \sigma_L$ (errors are statistical)

Table 2 Structure factor parameters (in Å) combined from the combined fits assuming smeared step function

Diblock	D Å	σ_D Å	N
5DH8HH (2%)	404 ± 5	245 ± 2	∞
5DD10HD (1%)	496 ± 10	253 ± 7	16 ± 4
6HH10DH (1%)	365 ± 7	233 ± 3	16 ± 4
6HH10DH (2%)	362 ± 5	227 ± 4	24 ± 4
10DH16HH (2%)	842 ± 12	424 ± 10	∞
6HH15DH (2%)	690 ± 8	442 ± 10	24 ± 4
6DH20HH (2%)	1043 ± 10	567 ± 8	24 ± 4

D repetition distance, σ_D smearing,
N number of platelets in the macro-aggregate

siderably smaller than that found in melt grown polyethylene crystals where the long period is smaller than 100 Å.

For the higher molecular weight diblock copolymers, stacking towards macro-aggregates always occurs. This is different for smaller molecular weight materials [8, 79]. Figure 20 presents the absolute cross-section for solutions of 1.5HH5DH PE-PEP diblocks under core contrast conditions. A comparison with Fig. 16 immediately shows that the scattering profiles differ qualitatively. At a concentration of 2% no platelet aggregation occurs and the observed stapling of platelets, which has been related to the van der Waals interactions between different platelets, obviously is absent. The data sets

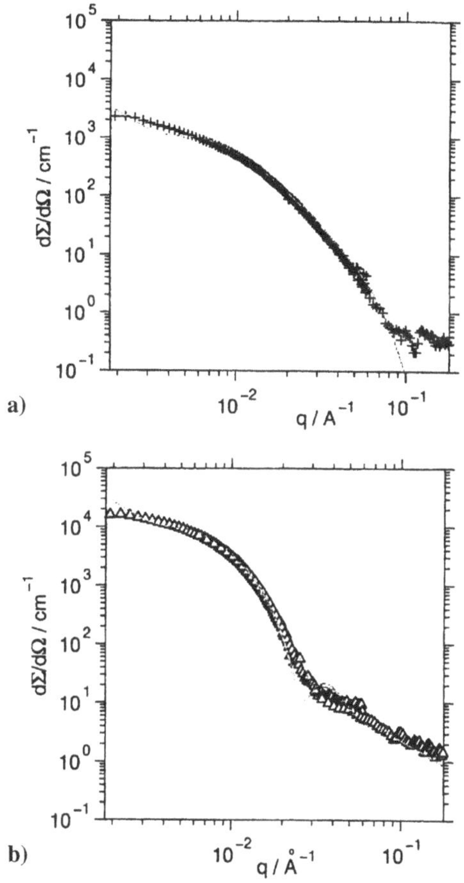

Fig. 20 **a** Scattering intensity from $\phi = 2\%$ solution of $M_w = 1.5$ K/5 K PE-PEP in decane (core contrast). **b** Brush contrast

under core (Fig. 20a) and brush (Fig. 20b) conditions display the characteristic differences. While the core intensity extends to $Q = 0.08$ Å^{-1}, the observed intensity under brush contrast drops rapidly at lower Q ($Q \cong 0.02$ Å^{-1}) indicating the much larger lateral extension of the brush compared to the core. Furthermore, above $Q \cong 0.02$ Å^{-1} the brush data display a power law tail originating from the scattering from the blobs. This power law tail is absent under core contrast.

An attempt to fit the data with the platelet form factors developed in Sect. 2.4 fails because the separation of length scales (platelet size much larger than lateral platelet extension) is not fulfilled. Therefore, the characteristic Q^{-2} power laws originating from platelet scattering are not visible. The actual size of the platelets influences the scattering pattern observed by the SANS instrument.

An attempt to describe the scattering data in terms of an average platelet size was not possible. Therefore, in a first approximation a model distribution of platelet sizes was used. The curves represented in Fig. 20 display the results obtained from platelet sizes of $r_1 = 200$ Å and $r_2 = 2000$ Å at a ratio of 0.35/0.65. The platelet sizes are rather ill defined and the numbers quoted above may be only used as orders of magnitude [8].

Information on the organization of the platelets within the macro-aggregates was revealed by SANS on diblock solutions under shear [7]. The application of shear in colloidal and polymer physics [80–85] is often used in order to induce ordering phenomena or to orient already locally ordered structures. Investigations were reported on the 6HH/15DH PE-PEP diblock copolymer at a volume fraction $\phi = 10\%$. Gel-like samples were brought between two thin quartz plates that were sheared relative to each other. Afterwards they were oriented in different ways relative to the incoming neutron beam. Figure 21 displays the qualitative changes of the two-dimensional scattering pattern when the sample is turned around and the axis becomes parallel or perpendicular to the shear direction. Figure 21a presents the geometry that would be equivalent to that of a Couette cell, revealing a scattering pattern with spots above and below the horizontal plane. Figure 21b shows the pattern arising if the sample is turned around an axis parallel to the shear direction by an angle of 45°. Under these circumstances no change compared to the outgoing situation of Fig. 21a is observed. In Fig. 21c and d the effect of turning around an axis perpendicular to the shear is displayed. Turning at 45° already smears out the pattern considerably. Increasing this angle to 85° finally reveals an isotropic pattern.

These qualitative observations allowed a clarification of the way the PE-PEP aggregates assemble within the macro-aggregates. The corresponding structures are shown in the inserts. As mentioned above, from observation with the phase contrast microscope we know that the macro-aggregates have needle-like shapes. Within these needles the planes of the PE lamella must orient parallel to the long direction of the needles; the needles themselves orient with their long direction parallel to the shear. Thereby the surfaces normal to the PE planes are oriented in a random fashion perpendicular to

Fig. 21 Shear experiments. The figures display schematically the scattering geometry, the shear direction, and the tilt of the sample, which stays between two thin quartz plates. The obtained scattering patterns are displayed on the *right side* of the figures, and the *inserts* in the *middle* visualize the platelet organization within the needle-like macro-aggregates. **a** Perpendicular geometry: neutron beam direction and shear direction are perpendicular. **b** Sample is turned by 45° around an axis parallel to the shear direction. **c** Sample is turned around an axis perpendicular to the shear direction by 45°. **d** Sample is turned by 85° around an axis perpendicular to the shear direction. All experiments were performed on the diblock 6HH15DH at $\Phi = 10\%$ and room temperature under core contrast ▶

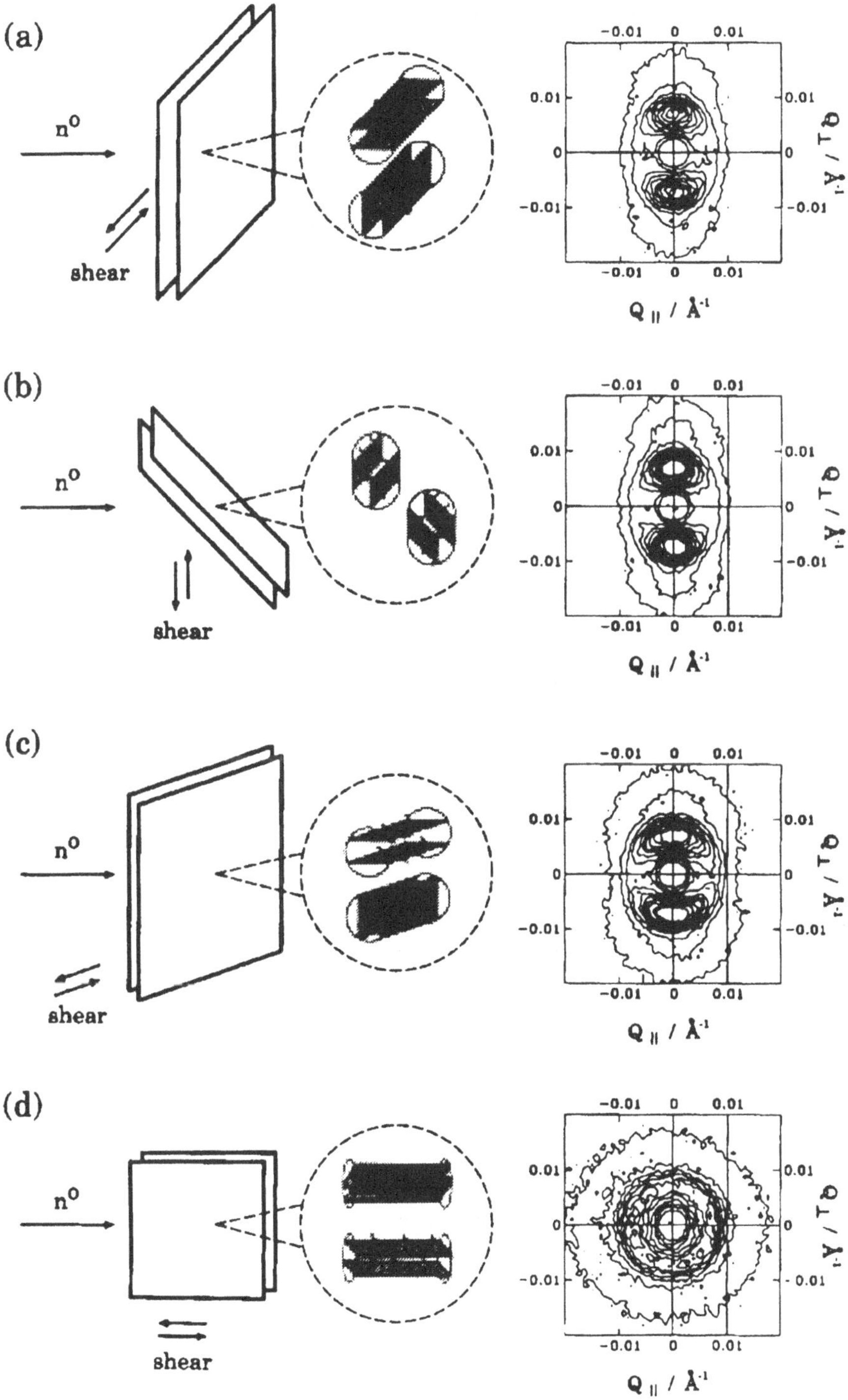
(a)
n°
shear
(b)
n°
shear
(c)
n°
shear
(d)
n°
shear
-0.01
0
0.01
Q ∥ / Å⁻¹
Q ⊥ / Å⁻¹

the needle direction. In this picture, the shear results are explained. Turning around an axis parallel to the shear direction brings no change. Out of the ensemble of PE planes there is always a fraction with an orientation fulfilling the Bragg condition. Those planes give rise to the reflexes above and below the scattering plane. The situation changes if the sample rotates around an axis perpendicular to the shear when more and more planes are brought into the Bragg condition. At 85° basically all planes, whatever their orientation within the needle, may give rise to Bragg scattering and produce a Debye–Scherrer ring.

4.3 Thermodynamics of Platelet Formation

The thermodynamics of platelet formation are governed by the gain of crystallization enthalpy opposed by the energetic penalty of the entropy increase due to chain stretching in the brush (Fig. 22) and the enthalpy of chain folding at the core surface. As outlined by Raphael and de Gennes [86] for triblock copolymers from flexible and stiff segments, the line tension at the platelet edge might limit the size. This phenomenon was not considered here. However, as will be seen, another important contribution comes from the defect energies of the ethylene side chains (which are present as the consequence of the anionic polymerization process).

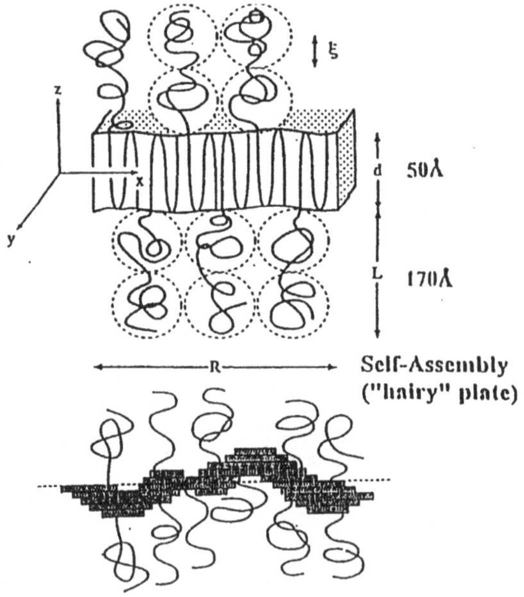

Fig. 22 Generic structure of the PE-PEP aggregates. The *lower part* indicates that the crystallization core exhibits some amount of surface roughness

The Alexander de Gennes [52, 53] brush model considers the entropic situation in a brush. It is derived from an analogy to semi-dilute polymer solutions. There the screening of the excluded volume interactions leads to a blob structure with swollen chain sections inside the blobs [87]. In this approach each hair may be considered as a string of blobs reaching out from the surface. The blob size describes the screening of the excluded volume interactions due to the neighboring chains. Its size relates to the surface density of the blobs Ω where:

$$\xi \approx \Omega^{-0.5} . \tag{47}$$

The hair surface density directly relates to the core thickness d and the chain length N_{PE} of the polyethylene subchain:

$$\Omega = \frac{0.5d}{V_0^{\mathrm{PE}} N_{\mathrm{PE}}} \tag{48}$$

where V_0^{PE} denotes the volume of a monomer in the PE subchain. Inside a blob the chains are swollen and we have:

$$\zeta \approx \ell g^{\nu} , \tag{49}$$

where l is the length of a monomer, g the number of monomers in the blob, and $\nu = 0.6$ is the Flory exponent. With the condition that the density inside one blob has to be equal to the average density, this consideration immediately leads to an expression for the brush length L_{P}:

$$L_{\mathrm{P}} = N_{\mathrm{PEP}} \left[\frac{0.5d}{V_0^{\mathrm{PE}} N_{\mathrm{PE}}} \right]^{1/3} \ell^{5/3} , \tag{50}$$

where N_{PEP} represents the number of monomers in the PEP segment. This derivation of L_{P} assumes a rectangular brush profile whereas the SANS results have shown that the actual profile is closer to parabolic or a smeared Gaussian. This, however, does not alter the scaling results.

We now turn to the free energy of the aggregate [88]. Two contributions need to be considered. The chain stretching within the brush leads to a loss of conformational entropy that can be estimated by the number of blobs (n_{b}) per chain. Each blob contributes $k_{\mathrm{B}}T$ to the free energy. Since $n_{\mathrm{b}} = L_{\mathrm{p}}/\xi$, with Eqs. 47, 48, and 50 we obtain:

$$F_{\mathrm{brush}} = k_{\mathrm{B}} T N_{\mathrm{PEP}} \ell^{5/3} \Omega^{5/6} . \tag{51}$$

The enthalpic part of the free energy contains two contributions. The first results from the chain folding and is simply given by the number of chain folds n_{f} times the fold energy of $E_{\mathrm{f}} = 170$ meV [89]:

$$F_{\mathrm{core}} = E_{\mathrm{f}} n_{\mathrm{f}} = \frac{2E_{\mathrm{f}} \ell N_{\mathrm{PE}}}{d} = \frac{E_{\mathrm{f}} \ell}{V_0^{\mathrm{PE}} \Omega} . \tag{52}$$

In addition to the fold energy there exists a further important contribution to the core free energy that results from the defect energy caused by the ethyl side branches. Gaucher and Seguela [90] found a defect energy of $E_{def} \cong 600$ meV. In addition they also found a strong tendency of the system to expel the ethyl branches into the amorphous phase. In other words, the ethyl branch frequency leads to a more frequent chain folding in order to avoid ethyl group incorporation into the bulk crystalline phase. For a low number of ethyl branches the Poisson distribution of side groups is very close to an equal distribution, yielding a rather uniform probability of $p = 0.017$ of finding an ethyl unit at any backbone carbon. An increase in the PE surface area serves to enhance the probability of locating an ethyl branch in the vicinity of the brush platelet surface. In the neighborhood of the surface the energy costs of incorporating a branch into the crystal is markedly diminished, i.e., in the region of a fold these energy costs are minimized. In this spirit an expression may be formulated for the defect contribution to the aggregate free energy:

$$F_{def} = [F^{\circ}]_{def} - n_f n_s p E_{def} = [F^{\circ}]_{def} - \frac{n_s p E_{def} \ell}{\Omega V_0^{PE}} . \qquad (53)$$

There $[F^{\circ}]_{def}$ is the defect energy for the incorporation of the ethyl branches into a large PE crystal, and n_s denotes the number of methyl units in the neighborhood of the fold at which the energy costs for ethyl group incorporations are strongly reduced. The product $n_f n_s \times p$ represents the number of ethyl units that do not contribute to the defect energy. We note that F_{def} scales as F_{core} with Ω^{-1}. Thus, the scaling properties of the total free energy are not affected by the addition of these defect terms. Finally, the total free energy may be written as:

$$F = \frac{\ell}{V_0^{PE}} (E_f - n_s p E_{def}) \Omega^{-1} + k_B T N_{PEP} \ell^{5/3} \Omega^{5/6} . \qquad (54)$$

Minimization of the total free energy with respect to Ω leads to the following scaling relations:

$$d = C_1 N_{PE} N_{PEP}^{-6/11} \qquad (55)$$

$$L_P = C_2 N_{PEP}^{9/11} \qquad (56)$$

where the prefactors C_1 and C_2 may be determined from the experiment. Figures 23 and 24 display the experimental test of the scaling relations Eq. 55 and Eq. 56. Both graphs show a reasonable agreement with the proposed scaling laws.

From Fig. 23 we obtain the proportional constant $C_1 = 6.7$ Å while Eq. 51 reveals $C_2 = 3.4$ Å. Inserting this result into Eq. 48 and introducing it into Eq. 54 we may estimate the fictitious core thickness for the case where ethyl side branches are absent. Then, the high fold energy E_f would require

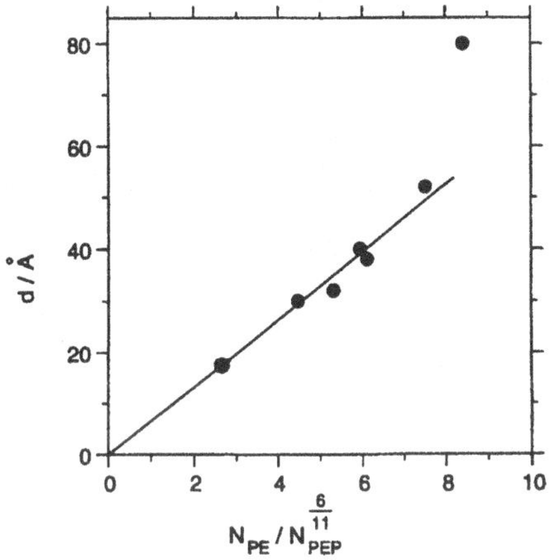

Fig. 23 Scaling plot of the core thickness, *d*, according to Eq. 53

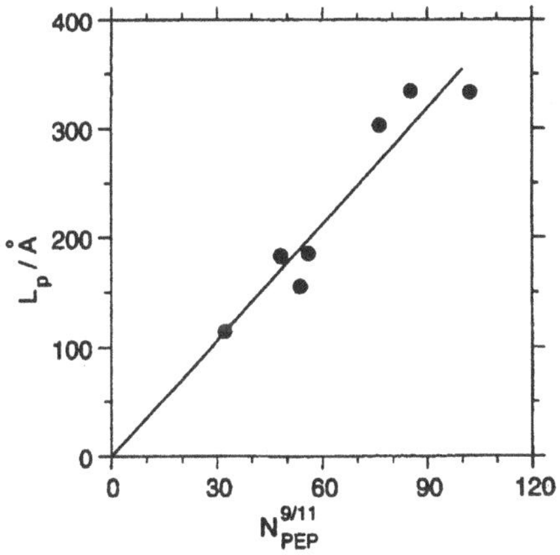

Fig. 24 Scaling plot of the brush length, L_p, according to Eq. 56

a much larger core thickness represented by a constant $C_1 = 26.4$ Å. Thus, the existence of the ethyl side branches favors thin platelets with a large surface area.

4.4 Interaction of PE-PEP Diblocks and Waxes

A display of the nucleating capacity of the PE-PEP diblock (1.5/5 K) is shown in Fig. 25. These micrographs were taken in a cold room at −13 °C (±1 °C) and represent a commercial fuel prior to and after the addition of a formulation involving the self-assembled crystalline-amorphous diblocks and a growth arresting copolymer. Obviously, the PE-PEP diblock aggregates are

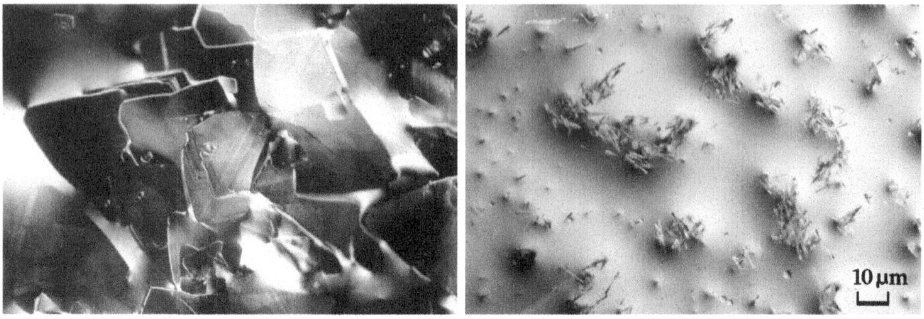

Fig. 25 Micrographs, at –13 °C, of an untreated (*left side*) and treated (*right side*) diesel fuel

able to suppress wax crystallization quite significantly. Since it is well known that the crystallization of waxes in solution proceeds via nucleation, it is suggestive to assume that the PE crystal surfaces of the diblock aggregates act as nucleation sites for the crystallization of waxes rather than to imagine cocrystallization of wax molecules with or within the PE crystalline core. We will later describe small angle scattering experiments that prove this assumption.

Under these circumstances the available platelet surface area (APSA) per gram of block copolymer in the fuel is the determining parameter. Considering the molecular weight of the polyethylene chains M_{PE} and the polyethylene density ζ_{PE}, the surface area per chain (SA) becomes:

$$\mathrm{SA} = \frac{2M_{PE}}{d\zeta_{PE}N_A}, \tag{57}$$

where d should be taken from Eq. 55 with the prefactor C_1 obtained from SANS measurements. From this surface we have to subtract the surface occupied by one hair:

$$S_{\mathrm{hair}} = \delta\frac{v_0^{\mathrm{PEP}}}{\ell} = \delta\frac{m_0^{\mathrm{PEP}}}{\zeta_{\mathrm{PEP}}\ell N_A}. \tag{58}$$

Here v_0^{PEP} is the volume of a PEP monomer, m_0^{PEP} is the monomer mass, ζ_{PEP} is the PEP density, and δ allows an effectively larger surface area than is blocked by the PEP hairs. Subtracting Eq. 58 from Eq. 57 yields the available surface area per chain. If we now multiply by the number of chains per gram we arrive at the required quantity, the available platelet surface area per gram [8]:

$$\mathrm{APSA} = \frac{\ell}{\zeta_{PE}N_a}\left(\frac{m_0^{\mathrm{PEP}}}{m_0^{\mathrm{PE}}N_{\mathrm{PE}} + m_0^{\mathrm{PEP}}N_{\mathrm{PEP}}}\right)\left(\frac{2N_{\mathrm{PEP}}^{6/11}}{C_1} - \frac{\delta}{\ell}\right). \tag{59}$$

At this point we consider monomers as being a section of the backbone with four methylene units having four carbons each in the backbone. This methodology relates a polyolefin structure to that of the parent polydiene in the diblock. The value of Eq. 59 is that it predicts the platelet surface area as a function of the diblock composition and the segment molecular weights. Samples are given in Table 3. Introducing the C_1 values with and without ethylene side branches ($C_1 = 6.7$ and 26.4 respectively) it is seen that the linear polyethylene segments yields noticeably lower APSA values than the corresponding segments containing ethyl branches.

Table 3 Representative molecular characteristics of PE-PEP crystalline-amorphous diblock copolymers

Polymer	6/10	1.5/5
M_{PE} ($\mathrm{g\,mol^{-1}}$)	6000 (±5%)	1500 (±5%)
M_{PEP} ($\mathrm{g\,mol^{-1}}$)	10000 (±5%)	5000 (±5%)
ζ_{PE} ($\mathrm{g\,mol^{-3}}$)	0.93 (±0.02)	0.93 (±0.02)
ζ_{PEP} ($\mathrm{g\,mol^{-3}}$)	0.85 (±0.01)	0.85 (±0.01)
ν_{PE}	0.38 (±0.02)	0.19 (±0.02)
m_{s}	0.82 (±0.06)	1.67 (±0.12)
$\rho_{\mathrm{PE}}(10^{10}\,\mathrm{cm^{-2}})$	-0.33 (±0.01)	-0.33 (±0.01)
$\rho_{\mathrm{PEP}}(10^{10}\,\mathrm{cm^{-2}})$	5.06 (±0.77)	5.06 (±0.77)
APSA ($\mathrm{m^2g^{-1}}$)	173 [a]/36 [b]	251 [a]/33 [b]

[a] Ethyl branches present
[b] Zero ethyl branch content

The mutual influence of wax and diblock copolymer aggregates was studied in a series of SANS experiments [8, 79]. The first set of data focused on the core and the brush scattering, respectively. For this purpose samples were investigated under contrast conditions where the wax and the brush, or both the wax and the core, were matched. Such matching conditions are achieved by choosing the proper mixtures of hydrogenated in deuterated solvent and hydrogenated in deuterated wax, respectively. Figure 26 presents SANS data from the 6HH10DH PE-PEP diblock copolymer under core contrast conditions at a polymer volume fraction of $\phi = 2\%$. This figure compares two scattering results, one from the pure diblock copolymer solution and the other from the same solution including $\phi = 0.5\%$ of the paraffin C_{36} as the wax fraction. The contrasts for both experiments were such that only the core is visible. In the region $Q \geq \sim 10 \times 10^{-3}$ Å^{-3} (relevant for the core scattering) both data sets agree completely. Obviously the wax does not influence the PE core size. Cocrystallization, which would have been an option, does not take place in solutions that were cooled from 70 °C.

Figure 27 displays data from the same sample now under brush contrast. Again the figure compares SANS data with and without an invisible wax

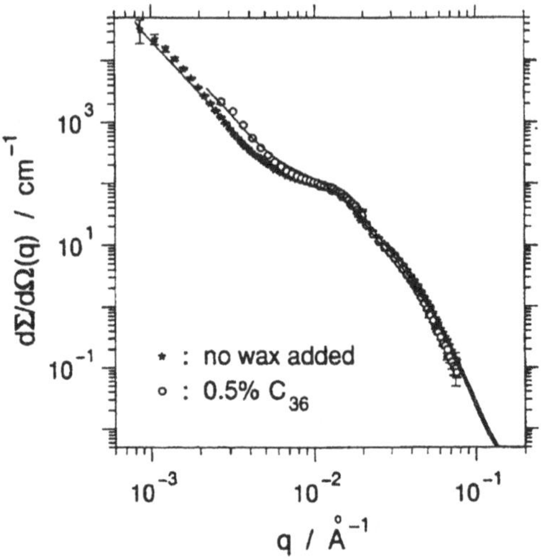

Fig. 26 Scattering cross-sections from the $M_w = 6\,K/10\,K$ diblock copolymer ($\phi = 2\%$) under core contrast, with and without the addition of 0.5% C_{36} wax having a scattering length density equal to the solvent, i.e., invisible

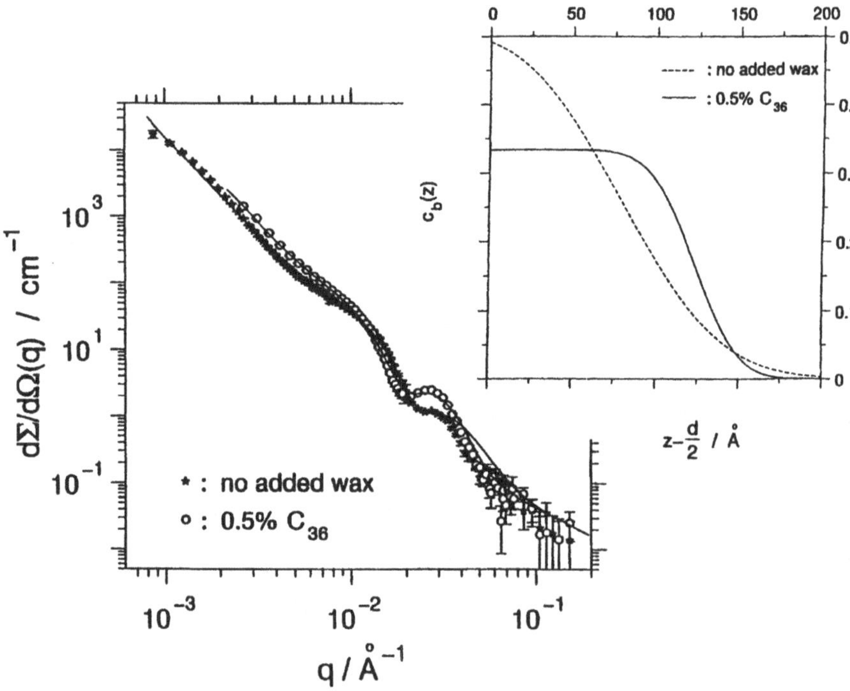

Fig. 27 Scattering of the same system as shown in Fig. 26 but under brush contrast. Core and wax are matched and therefore invisible. The *insert* shows the change of the brush concentration profile with and without wax addition

fraction of $\phi = 0.5\%$. Other than in the case of the core we realize that the brush scattering is modified. If wax is present the dip in the scattering pattern (indicating a zero in the form factor) moves to a smaller Q value. The subsequent secondary maximum moves accordingly and, in the case of the wax-containing sample, gains in structure indicating a sharper brush profile. The data were fitted with the brush form factor from Eq. 30 and the resulting profiles are displayed in the insert. With wax, the hair density profile is shifted outwards compared to the empty brush indicating more stretched hairs close to the surface. This effect is opposite to the influence of solved short chains in a long brush [91]. There these chains are distributed at the outside part and do not penetrate inwards. Furthermore, they cause brush compression. Obviously the crystallization enthalpy from the nucleation processes at the anchoring surface inverts the behavior. This observation may be taken as evidence that the paraffin associates itself with the PE surface and hence forces the hair profile away from the surface.

The association of paraffins with the polyethylene surface may be directly observed under a different contrast condition. If one chooses the contrast such that the hairs are matched but, at the same time, both core and paraffins have strong contrast an interaction of the wax with the PE surfaces will appear in the scattering experiments as an increase of the effective thickness of the core. For these experiments the H1.5HH5DH sample was chosen. This sample has a protonated core and the required contrast is achieved by taking protonated paraffins in a nearly fully deuterated solvent. In this experiment the effects of two different paraffins C_{36} and C_{30} were studied.

Figure 28 presents the scattering curves obtained at room temperature ($\phi_{\text{PE-PEP}} = 2\%$, $\phi_{\text{wax}} = 0.5$ or 0.62%). While the scattering profiles from the sample without wax and that from the system containing C_{30} coincide, a significant increase of the scattering intensity is observed for the sample containing the longer paraffin. From that we may directly conclude that at room temperature C_{30} does not aggregate at the platelets. On the other hand, the increase of the scattering intensity for the C_{36}-containing sample must relate to wax aggregation. The solid line presents a fit with the core form factor allowing for a variation of the effective core thickness. Assuming an average surface coverage by the wax from the fitted thickness d, the aggregated wax volume fraction ϕ_{wax} may be obtained. Assuming identical densities for the PE and the wax crystal we have:

$$\frac{\phi_{\text{PE}} + \phi_{\text{wax}}}{\phi_{\text{PE}}} = \frac{d_{\text{eff}}}{d} \,. \tag{60}$$

Using $\phi_{\text{PE}} = 0.0043$ and $\phi_{C36} = 0.0062$, $d_{\text{eff}} = 38$ Å and $d = 18$ Å we arrive at 86% aggregation of the C_{36} material.

An analogous evaluation of scattering data obtained at 5 °C reveals that the C_{30} wax is taken up by the PE-PEP aggregates, thus leading to a corresponding intensity increase. From the data an uptake of 57% of the added

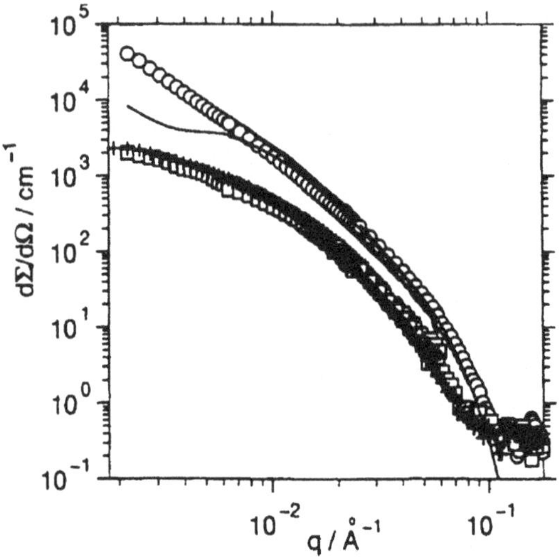

Fig. 28 Scattering at ambient temperature from 1.5HH5DH PE-PEP (2%) under core contrast (+), with addition of 0.5% C_{30} (□) or 0.62% C_{36} (○). The brush is matched, core and wax are visible. Only C_{36} shows a strong aggregation effect

C_{30} and nearly 100% of the C_{36} wax is accounted for. Thus, with decreasing temperature an increasing wax fraction is nucleated at the PE surface.

The SANS results have shown that the influence of the PE-PEP self-assembling diblock copolymer on wax crystallization originates from its nucleation properties at the active PE surface. The diblock copolymers suppress crystallization of waxes by providing a lower kinetic route to crystallization compared to normal nucleation of waxes in pure solutions. Theoretically it was shown that nucleating onto the PE core involves a lower energy barrier compared to crystallizing along the normal nucleation route. The specific rate of the process in the presence of the diblocks scales linearly with a number concentration of the crystalline-amorphous aggregate [92].

In practice, this relationship is well obeyed. Figure 29 displays results on the cold filter plugging point (ΣCFPP) for different diblock compositions plotted against the aggregate surface area. These surface areas were obtained from Eq. 59 using the appropriate diblock copolymer parameters. As may be seen, the cold filter plugging point data relates linearly to the surface area from the SANS results. The CFPP results are from the summation of data obtained from four different test fuels.

Finally, Fig. 30 displays schematically the crystalline-amorphous PE-PEP-wax aggregates. Polymer addition leads to the occurrence of many much smaller wax crystals than in the case of a pure wax solution, where the gel-forming house-of-cards arrangements of large platelet-like wax crystals are found.

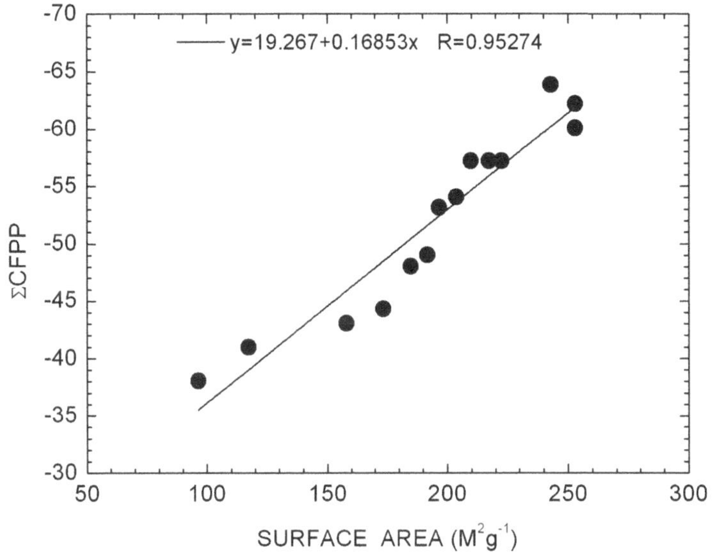

Fig. 29 Cold filter plugging point as a function of PE-PEP aggregate surface area (see text)

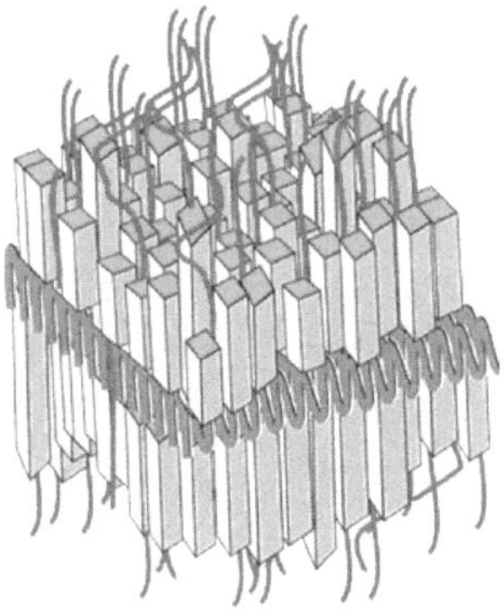

Fig. 30 Artists view of wax nucleation at the PE-PEP diblock aggregates

4.5
The Effect of PE-PEP Diblocks on the Yield Stress in Wax-Containing Oils

The efficacy of the PE-PEP diblock additives at preventing the formation of waxy gels in oil upon cooling below the wax solubility limit was studied by yield stress measurements in parallel with optical microscopy observations [15]. The effect of a low M_W diblock (1.5–5 K) on the yielding properties of waxy gels at a 4% wax level at 0 °C is shown in Fig. 31 for the different wax molecules considered. All gels display similar behavior with increasing additive concentration. At low polymer addition, the yield stresses of the gel drop

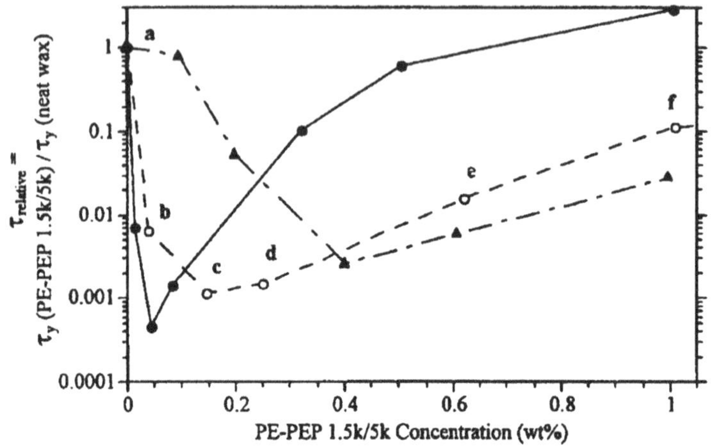

Fig. 31 Relative yield stresses (defined as the ratio of the yield stress of the gel with PE-b-PEP to that without the additive) of the 4 wt % wax (C_{28} through C_{36}) gels as a function of PE-b-PEP 1.5 K/5 K concentration at 0 °C. Mass fractions are reported on a total solution concentration basis. The *letters* correspond to the micrographs shown in Fig. 32. The yield stresses of the unmodified and 1 wt % PE-b-PEP 1.5 K/5 K modified C_{32} wax are measured at -5 °C. *Open triangles* C_{24}, *closed triangles* C_{28}, *open circles* C_{32}, *closed circles* C_{36}

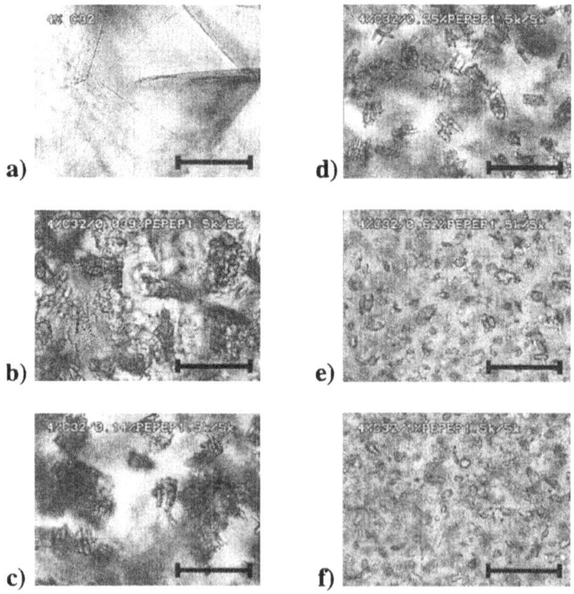

Fig. 32 Optical micrographs depicting the effect of PE-b-PEP 1.5 K/5 K on the morphology of 4 wt % C_{32} wax crystals in decane at 0 °C. The *solid bar* corresponds to a length of 100 μm. The figures (**a–f**) correspond to the similarly labeled points on Fig. 31

to a very low value in the order of more than thousand times lower than the yield stress in the unmodified gel. Further polymer addition beyond the minimum induces again an increase of the yield stress and leads to a re-gelling of the sample. While the general behavior is similar, the quantitative effects are nevertheless quite different.

Taking the polymer concentration at which the yield stress is at a minimum as a measure of polymer efficiency, it appears that the diblock copolymer aggregates appear to interact selectively with the different waxes; the lower the M_W, the stronger the effect on the longer chain waxes. Compared to the 1.5–5 K material, results from a 1.7–11 K PE-PEP diblock copolymer revealed that in this case the minimum yield stress of the wax gels is now shifted towards larger polymer concentrations by a factor of three. This result indicated that the diblock efficiency appeared to decrease with an increasing length of the brush. First it was speculated that the observed yield stress differences between the two materials could be attributed to the capacity to sequester the wax within the polymer brush layer, stabilizing the diblock aggregates in solution. This capacity may be evaluated as the brush volume available for paraffin crystallization before the aggregates are buried by wax. Following the considerations of Sect. 4.3 this capacity is given by:

$$\text{capacity} = \frac{\zeta_{\mathrm{PE}}\left(\frac{2L\nu_0^{\mathrm{PE}}N_{\mathrm{PE}}}{d-\nu_0^{\mathrm{PEP}}N_{\mathrm{PEP}}}\right)}{M_{\mathrm{PE}}N_{\mathrm{PE}}+M_{\mathrm{PEP}}N_{\mathrm{PEP}}}\,. \tag{61}$$

The thus-defined capacities for the two PE-PEP diblocks come out as Eq. 11 and 3.22 g-wax/g-polymer, respectively. This analysis would suggest that the polymer additive with the longer brush would be more effective at sequestering the wax and reducing the yield stress than the shorter diblock, contrary to experimental rheological evidence.

As shown by Fig. 29, the surface area is the important parameter. Calculating the available platelet surface area (APSA) defined by Eq. 59 for the two diblock copolymers, values of 292 $m^2\,g^{-1}$ and 213 $m^2\,g^{-1}$ are found. This is in accordance with the observation that the lower M_W material is more efficient in modifying the wax crystal structure.

Until now all considerations for explaining either the SANS or the rheological results were predicated on the argument that the polymer acts as a nucleation agent. This assumption rests on the fact that the PE-PEP assembly takes place at temperatures well above the wax crystallization point and is entirely true for the case of lower wax amounts or shorter paraffin chains like those considered in the SANS investigations. There the experiments have demonstrated that the PE core is unperturbed by the presence of the wax. However, for the special case of 4 wt % C_{36} wax in decane the wax cloud point is comparable to the PE-PEP aggregation temperature, around 40 °C. Therefore, steric stabilization of the wax crystals via cocrystallization could also take place. Cocrystallization events would place some fraction of the polymer on the surface of the wax crystal, preventing their fusion. It would account for the low amount of polymers that are needed to reduce the yield stress that the nucleation model alone cannot. Also it would explain why there is selectivity for a particular wax/polymer combination.

Further insights into the concentration dependence of the yield stress were obtained by examining the wax crystal morphology by an optical microscope. Figure 32 displays the morphological changes in the C_{32} wax crystal structure as a function of 1.5–5 K diblock concentration at 0 °C, following the points shown in Fig. 31. The unmodified C_{32} waxes crystallize in decane as large plate-like structures of about 100 μm size. Addition of the diblock copolymers with increasing concentration systematically reduces the crystal size such that the crystals at the largest concentrations are on the order of 10 μm in size.

For the images in Fig. 32b–d Brownian motion of the foreground crystals was observed relative to the out-of-focus background crystals. This observation is consistent with the diminished network strength observed rheologically. At the two highest polymer concentrations the network appears to be frozen (no Brownian motion was evident), an observation that relates favorably to the increasing yield stress observed.

5 Crystalline-Amorphous Poly(ethylene-butene) Copolymers

Until now we have discussed wax crystal modifiers that consist of crystalline-amorphous copolymers with the crystalline and amorphous sequences arranged either in a random (EVA materials) or orderly (PE-PEP diblocks) fashion. The structural and rheological studies concluded that control of the wax crystallization process is carried out by the nucleation of waxes via the polymers crystallizable segments. Also, for both types of wax modifiers, a selective trend for a particular wax polymer combination was observed. This behavior invited the evaluation of other polymeric architectures able to exhibit tunable crystallization tendencies in order to match the precipitation behavior of different waxes. One such family is represented by nearly random copolymers of ethylene and 1-butene. Like EVA these polymers consist of microcrystalline ethylene units copolymerized with amorphous segments. The microcrystallinity of the poly(ethylene-butene) random copolymers can be tuned by changing the ratio of ethylene to butene segments. A characterization of the sequencing statistics of branched units can be done by considering the reactivity ratio products (see Sect. 1.1).

We will first present an overview of the experimental findings in terms of "structure diagrams": temperature versus wax volume fraction maps that indicate the different structures found by SANS experiments for each of the investigated polymer–wax systems. This overview will be then followed by a detailed presentation of the structures and morphologies formed by the random copolymers and waxes in separate and in common solutions. First, we will consider the polymer self-assembly and report on SANS experiments on decane solutions of PEB-*n* copolymers with variable ethyl branch content, molecular weight, and polymer concentration [10, 12–15, 93]. Secondly, we will discuss the wax crystal modification properties of these polymeric systems from structural and rheological points of view. The ability of PEB-*n* copolymers to improve the cold flow properties of waxy oils was studied by characterizing the yield stresses of wax gels in decane doped with different PEB-*n* systems [12, 17], while the microscopic interactions of the polymer with the waxes was elucidated using neutron scattering. The contrast matching technique was used to identify separately the conformation and structure of the polymers and waxes for aggregates from common solutions [11–15, 93, 94]. The structure and morphology of the polymer, wax, and wax–polymer mixed aggregates have been typically studied by classical pinhole SANS instruments where the *Q* range covered (between 0.002 and 0.2 $Å^{-1}$) allowed an exploration of a length scale from 1 nm up to 100 nm. For some particular systems the hierarchical multilevel structures formed within a very wide length scale (up to 10 μm) were additionally investigated by USANS at a double-crystal diffractometer, within a *Q* range between 2×10^{-5} and 2×10^{-4} $Å^{-1}$ and by involving the focusing-mirror FSANS instrument in Jülich [38–40], which covered the *Q*-gap between the USANS and classical SANS (between 1×10^{-4} and 2×10^{-4} $Å^{-1}$). A special approach of the USANS and focusing-mirror SANS experiments was required due to the necessary slit geometry corrections of the USANS data [95] and the avoidance of multiple scattering effects [15, 95, 96]. Complementary to the USANS and FSANS investigations, optical and transmission electron microscopy observations were done for most of the samples examined. This provided useful additional information for a better understanding of the neutron scattering results.

5.1 Structure Diagrams

An overview of the experimental findings of the systematic SANS investigations of different combinations of PEB-*n* copolymers and wax molecules was done in terms of structure diagrams for each of the investigated polymer–wax mixed systems [12]. These diagrams display temperature versus wax volume fraction maps that indicate the different structures found by SANS experiments and reveal the wax and copolymer conformation within the joint formed structures (Figs. 33–35).

Each symbol stands for one experiment, while the symbol itself indicates the structure found. In all cases, we can distinguish between the polymer and the wax structures that are observed under the respective contrast conditions. The structural areas are characterized by the following symbols: *C* stands for coil conformation, *R* for rod- or needle-like structures, *P* for plates, *B* for three-dimensional bulk structures, *RS* for rods with structural modulation along the rod, and *PS* for structurally correlated plates. The dashed line in each diagram signifies the wax solubility line determined by observation of the CP [12], while the solid lines separate regions of different structures. Figure 33 displays the structural diagram for PEB-11 and C_{24} wax in decane and shows:

(i) Polymer structure: At temperatures higher than 0 °C polymer coil conformations are observed. Below this temperature and at low wax concentrations, rod-like precipitates start to appear. Below the wax solubility line, in all cases, polymer plates are found.

(ii) Wax structure: Above the solubility line, as expected, single wax molecules in coil conformation are found. Below and still slightly above the solubility line at low wax concentrations, wax plate structures prevail. For higher wax concentrations, wax bulk phases start to appear, seemingly crystallized from the polymer–wax plates.

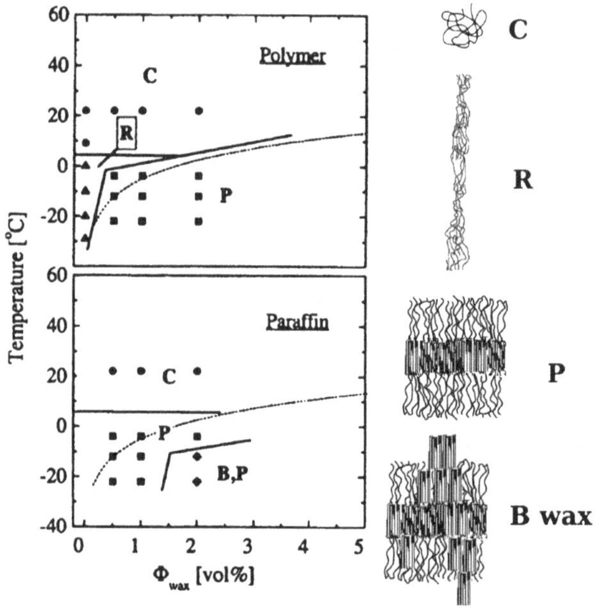

Fig. 33 Structures diagram for the PEB-11/C_{24} system in decane: the temperature dependence of the polymer and wax structure evolution as a function of wax content in solution. The *capital letters* denote the structural regimes delimited by lines and identified as: *C* coil conformation, *R* rod structure, *P* platelets, *B* three-dimensional bulk aggregates. On the *right side* the different structures are sketched schematically. The *data points* indicate the temperature/wax concentration conditions, where SANS experiments were performed, while the *symbols* denote the structures (*closed circles* coils, *closed triangles* rods, *closed squares* platelets and *closed diamonds* three-dimensional bulk aggregates)

Figure 34 shows the polymer wax structure diagram for PEB-7.5 and C_{24} wax in decane. Here the structure is more complex:

(i) Polymer structure: At temperatures above 45 °C single chain coil conformations are observed. In the regime between 20 and 45 °C apart from the polymer coils, large three-dimensional objects are found. Below 20 °C, still above the wax solubility line and at low wax concentrations, rod-like structures with density modulations along the rod axis are reported. At higher wax concentrations, still above the solubility line of the pure wax, polymer plate structures exhibiting interplatelet correlation appear. This structure covers the full low-temperature wax polymer regime.

(ii) Wax structure: Above 10 °C single wax molecules prevail. Below this temperature but already above the wax solubility line, there is some evidence for linear rod-like wax structures. In the regime where the polymer displays correlated platelet structures at higher temperatures and lower wax concentrations, the same structure is also observed for the wax. For low temperatures and higher wax concentration, wax bulk phases are present. These originate from primary polymer wax cocrystallized platelets and

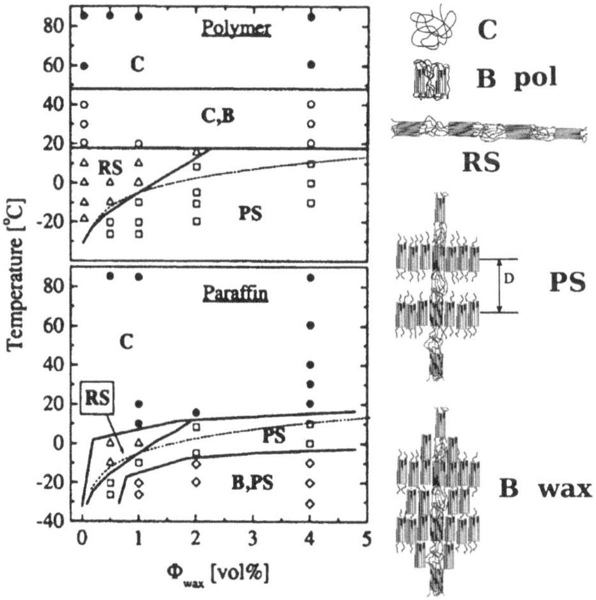

Fig. 34 Structures diagram for the PEB-7.5/C_{24} system in decane. *Lines* and *symbols* have the same meaning as in Fig. 32. In addition, *B-pol* denotes polymeric bulk aggregates; *RS* rods with internal (modulated) structure (*open triangles*); *PS* structurally correlated platelets (*open squares*), which grow laterally from the primordial modulated rod structure; *open circles* coexistence of coils and three-dimensional bulk aggregates; *open diamonds* coexistence of structurally correlated plates and three-dimensional objects. The different structures are sketched on the *right side* of the figure

represent the thick wax platelets observed by FSANS which, by continuous wax incorporation, transform into the large scale one-dimensional aggregates revealed by microscopy.

Figure 35 presents the structure diagram for the PEB-7.5 (6 K) and $C_{36}H_{74}$ mixtures where the most complex variety of structures was observed:

(i) Polymer structure: At high temperatures the polymer coils coexist with polymer plates formed under the wax influence. For zero wax concentration below 45 °C, in addition to the single coils, three-dimensional bulk structures appear. At 0 °C and below, these aggregates coexist with the rod-like polymer structures presenting density modulation. At low wax content and moderate temperatures (20–40 °C), three-dimensional bulk aggregates and platelets appear to coexist. At lower temperatures and higher wax concentrations, plates dominate the entire temperature–wax concentration regime. Further, at high wax concentrations three-dimensional polymer structures coincide with plates.

(ii) Wax structure: At high temperatures, above 30–45 °C depending on wax content, we see the coexistence of free wax molecules in solution and wax plate structures. In a small temperature interval that narrows toward higher wax contents the plate structures dominate, while at low temperatures and high wax concentration, plates and bulk structures coexist.

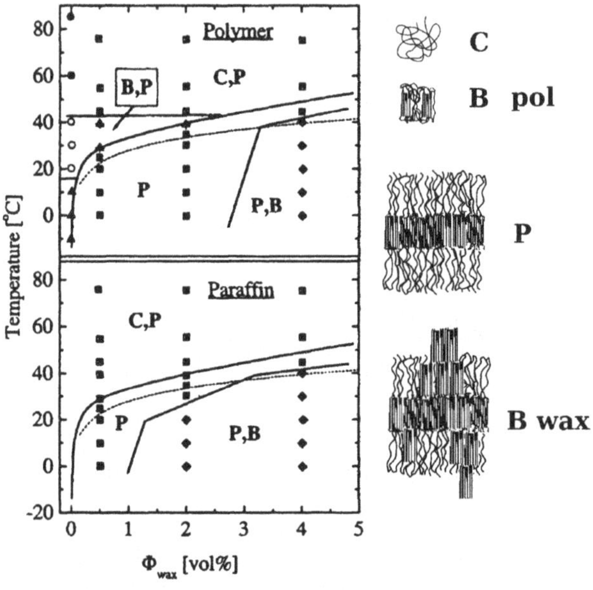

Fig. 35 Structures diagram for the PEB-7.5/C36 system in decane. *Lines* and *symbols* have the same meaning as in Fig. 33. In addition, *crossed squares* denote the coexistence of coils and platelets, and *crossed triangles* the coexistence of platelets and three-dimensional aggregates. The different structures are sketched on the *right side* of the figure

The structure diagrams offer an easy and direct way to watch the polymer influence on the shape and, in connection with the parameters yielded by the SANS results, the size of the wax crystals as a function of temperature and different wax contents. An inspection of these maps shows that, for the wax content and temperature interval of direct interest for the technical problems occurring in transportation of crude oils and use of middle distillates during winter time, the PEB-*n* copolymers represent efficient wax modifier additives. They can control the wax crystallization such that a deterioration of the viscoelastic properties of oils and fuels as a consequence of large waxy crystals is prevented. Instead of gelling into house-of-cards agglomerates hundreds of micrometers in size, small two-dimensional crystals or one-dimensional bulky structures fulfilling the size criteria required by the CFPP parameter are formed. Obviously, these modified morphologies worsen to a far lesser degree the viscoelastic properties of oils and fuels, and are much less able to induce a worsening of the viscoelastic properties of oils and fuels than the untreated cases.

5.2 Self-assembling of Random Crystalline-Amorphous Copolymers (PEB-*n*)

The aggregation features of the PEB-*n* copolymers were examined as a function of M_W and polymer concentration by investigating fully protonated copolymers in *d*-decane over the temperature range between the single coil and the aggregates regime (85 to –30 °C). Two representative polymeric systems were the subject of neutron scattering studies, the less crystalline PEB-11 and the highly crystalline PEB-7.5. The characterization data are given in Table 4.

The self-assembling behavior of PEB-11 (6 K) and PEB-7.5 (6 K) can be deduced from the scattering profiles displayed in Fig. 36. Decreasing the temperature leads to a gradual increase of the scattering intensity at low Q for both polymer solutions, indicating more pronounced aggregation of polymers at lower temperatures (Fig. 37). For PEB-11, the copolymer with the higher density of ethyl side groups, the single chain regime extends over a broad range of temperatures while, for PEB-7.5, the copolymer with the higher crystallization tendency, the single chain regime is found only for a limited range at high temperatures. Obviously, the aggregation process for each of the polymers in decane starts at different temperatures that are lower the higher the ethyl content is. While PEB-7.5 shows a tendency of slight but continuous aggregation already between 60 °C and 40 °C, PEB-11 starts to aggregate only below 0 °C. The scattering cross-section of both PEB-11 and PEB-7.5 polymers in the low temperature aggregation regime is characterized by a well-defined Q^{-1} power, which is indicative for rod-like aggregates. The size of the objects is not accessible by classical SANS since no saturation of the intensity toward small Q is visible in Fig. 36. Thus, the length scales of the

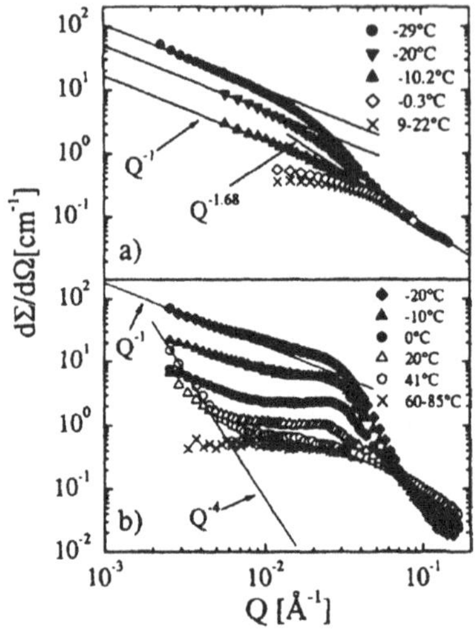

Fig. 36 Scattering cross-sections from 1% solutions of PEB-11 (**a**) and PEB-7.5 (**b**) random copolymers in *d*-decane. The *solid lines* indicate power law relations in different Q ranges

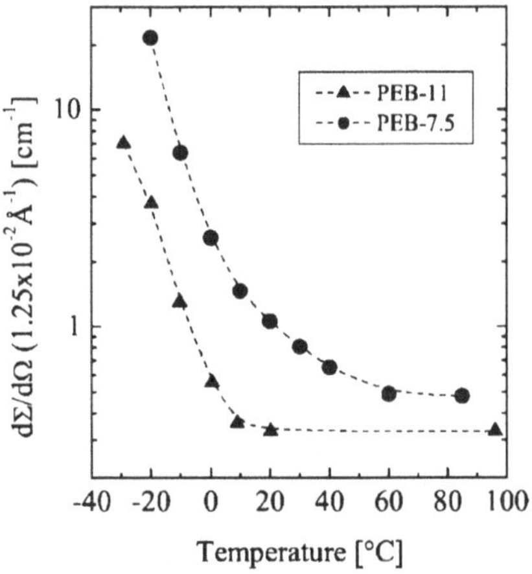

Fig. 37 Absolute cross-section from the PEB-11 and PEB-7.5 solutions at $Q = 1.25 \times 10^{-2}$ Å^{-1} as a function of temperature. The *dashed lines* serve as a guide for the eye

aggregates need to be larger than 100 nm in order to be compatible with the low-Q observations.

From the examination of the scattering patterns of PEB-11, the tendency toward a $Q^{-1/\nu}$ asymptote with the Flory exponent $\nu = 0.6$ observed at high Q appears to have only a very minor temperature dependence. Thus, though the polymers are aggregating, on a local scale the diffuse scattering from the monomer correlations within a swollen chain prevails, indicating very open rod structures with single chain structures on a local scale. The devi-

Table 4 Molecular characteristics and scattering properties of PEB-n copolymers

	hPEB-7.5	dPEB-7.5	hPEB-11	dhPEB-11	$C_{24}H_{50}$	$C_{24}D_{50}$	$C_{36}H_{50}$	$C_{36}D_{50}$	h-22	d-22
M_w (g mol^{-1})	6000	6000	6400	6400	338	388	506	580	142	164
M_w/M_n	1.02	1.02	1.02	1.02	1	1	1	1	1	1
ξ (g cm^{-3})	0.863	0.986	0.856	0.920	0.80	0.92	0.80	0.92	0.73	0.84
M_0 (g mol^{-1})	16.2	18.55	17.2	18.49	338	388	506	580	142	164
V_0 (cm^{3}mol^{-1})	18.8	18.8	20.08	20.08	422.5	422.5	632.5	632.5	194.5	194.5
ρ (10^{10} cm^{-2})	−0.31	7.42	−0.31	3.72	−0.39	7.04	−0.36	7.00	−0.49	6.58

ation from the Q^{-1} law at higher Q values yields information on the lateral size of rods. Within the Guinier approximation, the cross-section for one-dimensional aggregates has the form:

$$\frac{d\Sigma}{d\Omega}(Q) = \Phi_{\mathrm{rod}}(1 - \Phi_{\mathrm{rod}})\pi F_{\mathrm{rod}}\Delta\rho_{\mathrm{rod}}^2 \exp(-a^2Q^2/4)/Q\,, \tag{62}$$

where $F_{\mathrm{rod}} = \pi a^2$ is the perpendicular area of the rod, a the rod radius, Φ_{rod} is the volume fraction of rods, and $\Delta\rho_{\mathrm{rod}}$ is the contrast between the solvent and the rod region (the average over the polymer and solvent content has to be taken since the rods contain solvent).

If one plots $ln(Q\,d\Sigma/d\Omega)$ vs.Q^2, then such a one-dimensional Guinier representation directly reveals the rod radius (Fig. 38); pronounced linear regimes allowing an evaluation of the lateral rod size and forward scattering are observed. The resulting values for the rod radii and the "forward scattering" ($d\Sigma/d\Omega$ ($Q = 0$)) are listed in Table 5 for two volume fractions of PEB-11 in *d*-decane.

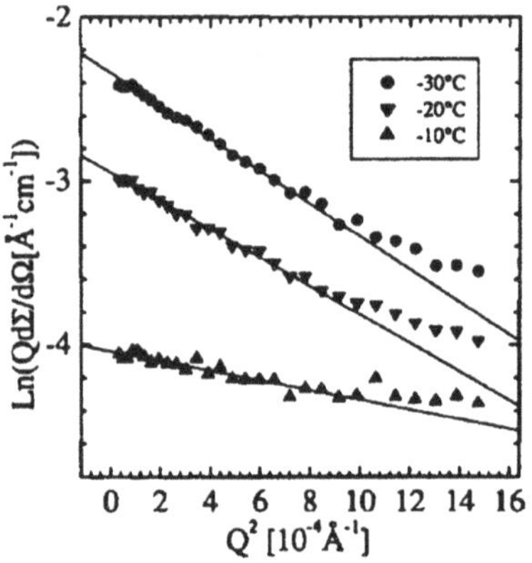

Fig. 38 Scattering patterns from 1% solution of PEB-11 in *d*-decane in a one-dimensional Guinier presentation

Table 5 Rod radius and "forward scattering" from the rod-like aggregates formed by PEB-11 in *d*-decane

Temp.	$\Phi_{\mathrm{pol}} = 1\%$		$\Phi_{\mathrm{pol}} = 2\%$	
	a	$Q\,d\Sigma/d\Omega\|_{Q=0}$	a	$Q\,d\Sigma/d\Omega\|_{Q=0}$
°C	Å	10^6 cm^{-2}	Å	10^6 cm^{-2}
−10	42 ± 2	1.76 ± 0.02	37 ± 3	0.55 ± 0.01
−20	72 ± 1	5.2 ± 0.1	82 ± 2	12.6 ± 0.1
−30	78 ± 1	9.6 ± 0.1	83 ± 1	23.0 ± 0.1

We deal with a partition of soluble polymers between the self-assembled aggregates and the remaining single chains in solution. From the evaluation of the forward scattering it is not possible to give exact values for the volume fractions of the samples occupied by the aggregated polymers and for the polymer density within the aggregated structures. However, both quantities are related to each other and this relation may be quantified taking into account the values of the forward scattering reported in Table 5.

We now use the following definitions of polymer volume fractions: Φ_{pol}^{agg} polymer volume fraction of the polymer in the aggregate; Φ_{pol}^{sol} polymer volume fraction of the polymer in solution; Φ_{rod} aggregate volume fraction in the sample; ρ_{rod} aggregate scattering length density; ρ_S solvent scattering length density, and ρ_P polymer scattering length density. The total amount of polymer in the sample Φ, aggregated and still in solution, stays constant:

$$\Phi_{pol}^{agg}\Phi_{rod} + \Phi_{pol}^{sol}(1 - \Phi_{rod}) = \Phi \,. \tag{63}$$

If we rearrange this equation we obtain:

$$\Delta\Phi \equiv \Phi_{pol}^{agg} - \Phi_{pol}^{sol} = \left(\Phi - \Phi_{pol}^{sol}\right) / \Phi_{rod} \,. \tag{64}$$

On the other hand, if $\Delta\rho_{rod}$ is expressed as $\Delta\rho_{rod} = \Delta\Phi\Delta\rho$, with $\Delta\rho = \rho_P - \rho_S$ known from the constitution of the molecules and the tabulated values of atom scattering lengths, Eq. 62 transforms into:

$$\Delta\Phi = \left\{\sqrt{Q\frac{d\Sigma}{d\Omega}(0) / \left(\pi F_{rod}\Delta\rho^2\right)}\right\} / \{\Phi_{rod}(1 - \Phi_{rod})\} \,. \tag{65}$$

We first evaluate the difference between the aggregated and free polymer volume fractions $\Delta\Phi$ as a function of Φ_{rod}, according to Eq. 64 for different Φ_{pol}^{sol}. Then following Eq. 65 and using the experimental values for $Q(d\Sigma/d\Omega)$ and $F_{rod} = \pi a^2$ from Table 5, Fig. 39 displays how the polymer volume fractions Φ_{pol}^{agg} and Φ_{pol}^{sol} in the aggregates and solution relate to each other as a function of the volume fraction of the formed aggregates $\Phi_A = \Phi_{rod}$. For a maximum possible volume fraction $\Phi_{rod} = 30\%$ all polymers would be inside the aggregates with the volume fraction of polymers in solution equal to zero and the polymer volume fraction inside aggregates retaining a very low value, $\Phi_{pol}^{agg} = 6\%$. If the aggregates would fill smaller volume fractions, the polymer volume fractions inside the aggregates would increase up to a maximum value of about 20%. This rationalization agrees with the observation of the scattering profile at high Q, showing that the aggregates represent open objects.

Other than PEB-11, in addition to the low temperature Q^{-1} power law behavior, a correlation peak evolves at around $Q = 0.025$ Å^{-1}. For the intermediate temperature range (between 40 and 20 °C), a Q^{-4} behavior at low Q appears. The low-Q Porod scattering indicates formation of large aggregates featuring a well-defined surface, while its attenuation with decreasing tem-

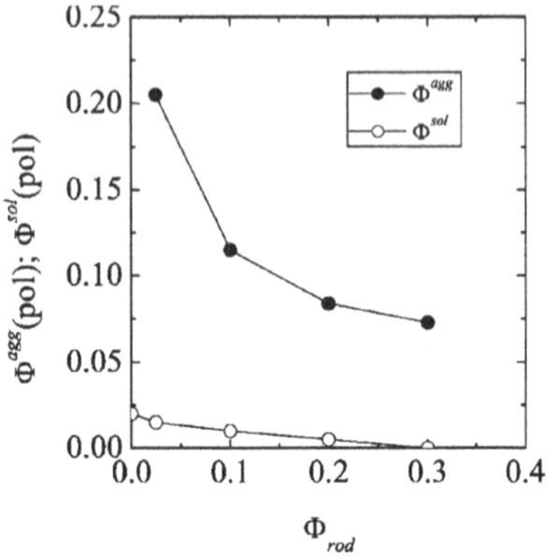

Fig. 39 Aggregated and solved polymer volume fractions $\Phi_{\mathrm{pol}}^{\mathrm{agg}}$ and $\Phi_{\mathrm{pol}}^{\mathrm{sol}}$ for PEB-11 as a function of the aggregates fraction Φ_{rod}

perature results from coarsening processes that remove the scattered intensity of the growing aggregates out of the observation window of SANS. FSANS measurements (Fig. 40) revealed that at Q values lower than those accessible by classical SANS, the scattered intensity increases dramatically as a Q^{-4}

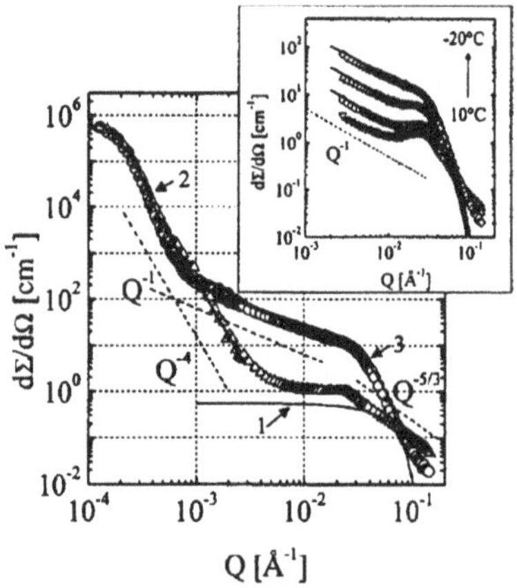

Fig. 40 Scattering patterns from a solution of PEB-7.5 copolymer (6 K) in *d*-decane at two temperatures within the aggregation regime: 20 °C (*open triangle*) and –20 °C (*open circles*). The *dashed lines* indicate the power law regimes identified in different Q ranges while the *solid lines* represent the model description of the data within particular Q ranges and for different temperatures. 1 The single coil form factor (Beaucage model for $P = 5/3$); 2 the form factor of large scale compact aggregates (Beaucage model for $P = 4$); 3 the density-modulated rods model Eqs. 32–34. The *inset* presents the temperature evolution of the main features characterizing the density-modulated rods, namely the Q^{-1} power law and the correlation peak; the *solid lines* represent model description of the data according to the density-modulated rods model

power law terminating in a Guinier-like regime around 2×10^{-4} Å^{-1}. This observation shows the formation of compact aggregates with sizes within the micrometer range. The temperature-independence of the low-Q scattering indicates that these large-scale aggregates neither grow in size nor number on decreasing the temperature below 20 °C.

Optical microscopy offers a direct visualization of these large compact objects at room temperature (Fig. 41). A close inspection of these objects (inset of Fig. 41) reveals that they show a dumbbell-like shape rather than a spherical one. The model interpretation of the scattering from dumbbell morphology [97] would reveal the scattering of neighboring spheres at low Q values while, within the Q range covered by FSANS, the scattering is governed by the individual spheres. Thus, a rough characterization of these objects was possible by combining the structural features observed in the micrographs with the parameters obtained from the interpretation of the FSANS data with Eq. 20 (Beaucage model). With the evaluated forward scattering and a sphere diameter of 2 μm, as revealed by the micrograph, it was concluded that a very small fraction (0.55%) of all polymers resides inside these aggregates. Formation of such large compact structure relates to the faster crystallization of those chains containing long and uninterrupted methylene repeat units.

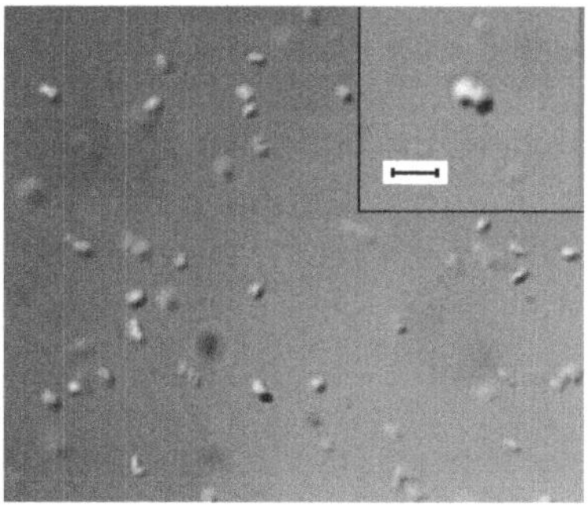

Fig. 41 Optical micrograph from a decane solution of 6 K PEB-7.5 copolymer at room temperature; the *scale bar* for the inset represents 4 μm

The peak-like feature that evolves on decreasing temperature at intermediate Q values (see also inset of Fig. 40) could denote either interparticle or intraparticle correlations. In the first case it would signify a structure factor arising from interaction effects between different rods, while for the latter the correlation would come from a density modulation (i.e., alternating crystalline-amorphous sequences) along one rod axis. The coincidence of the peak position for different polymer concentrations at low temperature (Fig. 42) led to the conclusion that the correlation length scale is not affected by the number density of the aggregates (to give rise to

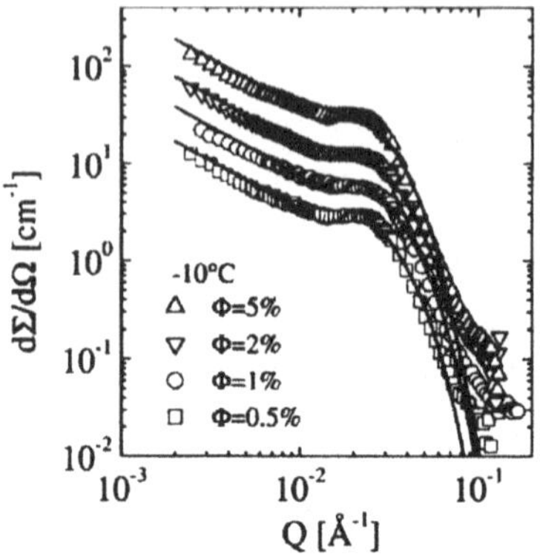

Fig. 42 Scattering patterns from PEB-7.5 (6 K) solutions in *d*-decane at −10 °C for different polymer concentrations; the *solid lines* represent the fit according to the density-modulated rods model

an interparticle structure factor). The conclusion was that it arises from a modulation of the density along the rod. The form factor appropriate for an ensemble of isotropically oriented cylinders with longitudinal density modulation is given by Eqs. 32–34. Figures 40 (inset) and Fig. 42 show that the experimental scattering patterns measured for a constant polymer content at different temperatures and for different polymer contents at a constant temperature are well described by this density-modulated rod model.

Table 6 contains the obtained structural and density parameters. Similar to the case of the PEB-11 copolymer, due to the partition of the dissolved polymers between the aggregates and the remaining single chains in solu-

Table 6 Parameters of the polymeric rods with modulated density formed by PEB-7.5 (6 K) in *d*-decane as obtained from the fitting of the experimental data with Eqs. 32–34

Φ_{pol}	Temp.	a	h	$\Phi_{rod}\Phi_{pol}^{agg}/\Phi_{pol}$ for max Φ_{pol}^{agg}	for min Φ_{pol}^{agg}
	°C	Å	Å	%	%
0.01	0	52 ± 1	310 ± 7	2.1	10
0.01	−10	53 ± 1	330 ± 5	6	30
0.01	−20	53 ± 1	370 ± 10	16.5	80
0.005	−10	57 ± 2	340 ± 10	5.5	26
0.02	−10	53 ± 1	340 ± 7	6	30
0.05	−10	60 ± 1	339 ± 7	4.6	23

a rod radius, *h* modulation length, $\Phi_{rod}\Phi_{pol}^{agg}$ volume fraction of the aggregated copolymer

tion, from analysis of the forward scattering no unique values for the volume fraction of rods nor for the polymer volume fraction within the rods could be obtained, but a relation between these quantities could. Thus, depending on the actual value of the volume fraction of rods in the sample, the polymer density is between 4% and 20%, which again proves the formation of rather open aggregates. Solutions of high M_w PEB-7.5 (30 K) random copolymers [14] revealed the same aggregation features as those of low M_w copolymers with the only difference being that the self-assembly occurs at higher temperatures. The model evaluation with Eqs. 32–34 resulted in a correlation length varying between 530 and 580 Å between room temperature and 0 °C and a potential maximum polymer density inside the rods of 36%. Obviously, there is no correspondence between the correlation length and the chain length of the stretched copolymers, which are about 550 and 2750 Å for the low and high M_w copolymers, respectively. Apparently, the modulation density along the rod does not relate directly to the chain length.

The main conclusion of the SANS studies was that the PEB-*n* random copolymers have a tendency to aggregate into rod-like structures. Although this effect is not fully understood, the following explanation was put forward as to why rod structures instead of loose three-dimensional networks are preferred. Both PEB-11 and PEB-7.5 are copolymers with segments of linear polyethylene modulated by branching with ethyl side groups. The PE blocks have different lengths and are statistically distributed along the chains. It may be assumed that the aggregation occurs as a consequence of crystallization of the PE sections. A crystallization process would promote in a first step the crystallization of PE segments within one polymer chain, while the amorphous parts forming a loose corona around the crystalline nucleus would lead to the screening of the crystal against the intrusion of other chains by osmotic repulsion. The rare cocrystallization events happening would lead to the preferential linear structures.

This picture is supported by SANS observations made in the case of the self-assembling behavior of two star polymers built from crystallizable amorphous PE-PEP diblocks. For one star copolymer the crystallizable blocks are placed in the core (PEP-PE) while the second star contains an amorphous core and a crystallizable corona (PE-PEP). The single star form factor in *d*-decane was studied at high temperature (70 °C) and for low star volume fractions ($\Phi = 0.24$) in order to avoid the self-assembling events and structure factor effects. Figure 43 shows the measured cross-sections for the two stars considered. The scattering profiles present the typical features for star scattering: (i) a low-Q Guinier regime; (ii) an intermediate-Q strong intensity decrease where the scattering regime changes from the coherent to the incoherent superposition of the arm scattering; and (iii) a high-Q tail characterized by a power law with the Flory exponent $\nu = 3/5$, which denotes excluded volume interactions between single chains. The cross-section for

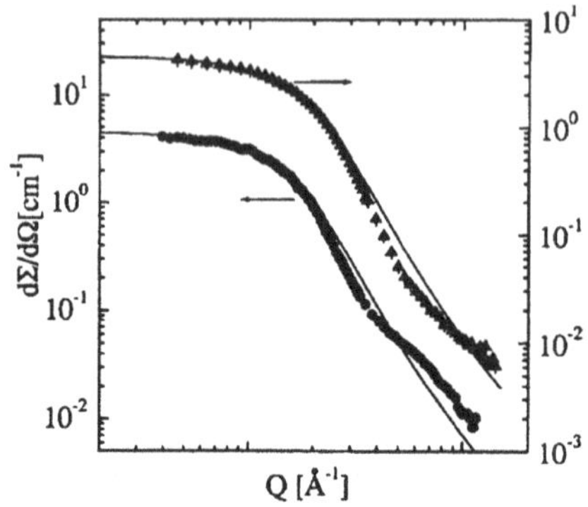

Fig. 43 Single star form factors at 71 °C for the PE-PEP and PEP-PE stars in *d*-decane for a polymer concentration $\Phi_{\mathrm{pol}} = 0.24\%$. The *solid lines* represent a fit with the Gaussian star form factor

single stars may be expressed as:

$$\frac{\mathrm{d}\Sigma}{\mathrm{d}\Omega}(Q) = \left(\Delta\rho^2/N_A\right)\Phi V_{\mathrm{arm}} f_{\mathrm{arm}} P(Q)\,, \tag{66}$$

where $\Delta\rho^2$ is the contrast between polymer and solvent, Φ the polymer volume fraction, V_{arm} the molar volume of one arm, f_{arm} the number of arms, and $P(Q)$ the star form factor [98, 99], which for Gaussian chains assumes:

$$P(Q) = 2\left\{u^2 - \left[1 - \exp\left(-u^2\right)\right] + (f-1)\left[1 - \exp(-u^2)\right]^2/2\right\}/(fu^4)\,, \tag{67}$$

with $u^2 = Q^2\left(R_g^2\right)_{\mathrm{arm}}$. A fit with Eq. 67 (solid lines in Fig. 43) describes well the low-Q behavior and the intermediate-Q fast decay of the scattering profile. At high-Q, due to the assumption of Gaussian chain statistics, the fit shows significant deviations from the scattering profile. The obtained fit parameters are presented in Table 7.

Figure 44 presents the SANS patterns from the PE-PEP star aggregates formed at −21 °C in *d*-decane for three polymer volume fractions. The scattering patterns reveal a well-defined Q^{-2} power law, which is characteristic

Table 7 Single star properties from SANS

	PE-PEP star	PEP-PE star
$\mathrm{d}\Sigma/\mathrm{d}\Omega/\Phi$ (cm^{-1})	1860	1880
$V_{\mathrm{W,arm}}$ (cm^3 mol^{-1})	7520	6050
$R_{\mathrm{g,arm}}$ (Å)	56.9	68
f_{arm}	31.5	31.7
R_{g} (Å)	97.6	116

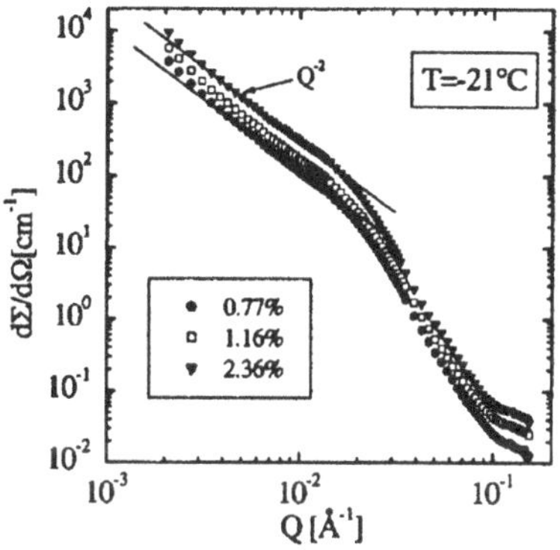

Fig. 44 Scattering patterns from the PE-PEP stars at – 21 °C for different polymer volume fractions. The straight lines depict the low-Q Q^{-2} power law behavior of the scattered intensity

for the scattering from platelets. In the Guinier approximation the scattering cross-section for two-dimensional aggregates has the form:

$$\frac{d\Sigma}{d\Omega} = \Phi_{\text{plate}} \Delta\rho^2_{\text{plate}} 2\pi d_{\text{eff}} \exp\left(-Q^2 d^2_{\text{eff}}/12\right)/Q^2 , \tag{68}$$

where Φ_{plate} is the volume fraction of the platelets, $\Delta\rho_{\text{plate}}$ the corresponding scattering length density contrast, and d_{eff} the effective thickness of platelets. Table 8 contains the parameters obtained for the self-assembled star aggregates. Obviously, these platelets contain crystalline as well as amorphous star sections and, thus, they incorporate a certain fraction of d-decane, which may be obtained from the absolute intensities supposing all stars are aggregating at this temperature (see Table 8).

Table 8 Parameters of the self-assembled PE-PEP star-based two-dimensional aggregates at –21 °C

Φ %	d_{eff} Å	$Q^2\, d\Sigma/d\Omega\|_{Q=0}$ 10^{14} cm^{-3}	$\Delta\rho_{\text{plate}}$ 10^{10} cm^{-2}	$\Phi^{\text{plate}}_{\text{pol}}$
0.77	136.2 ± 0.8	1.59 ± 0.02	3.51	0.51
1.16	140.8 ± 0.7	2.42 ± 0.02	3.45	0.50
2.36	147.6 ± 0.8	4.78 ± 0.02	3.17	0.46

Figure 45 presents the temperature-dependent SANS results measured in the case of 0.24% of PEP-PE stars in d-decane. Other than for the stars with crystallizable blocks at the corona rim, the scattering patterns from the PEP-PE aggregates do not change below room temperature and behave like Q^{-1} at low-Q. These stars aggregate into rods or needles, which involves all polymers

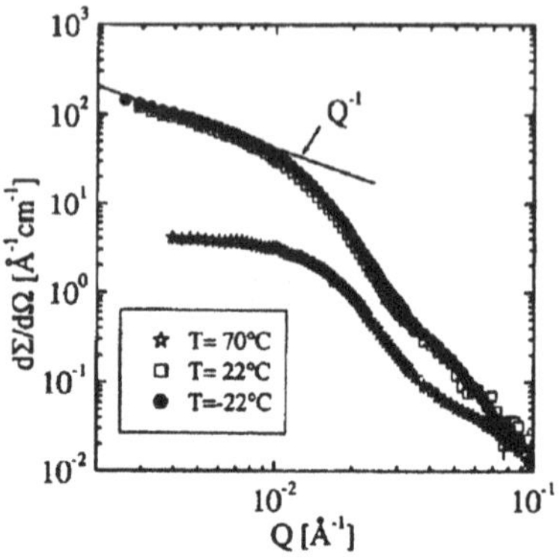

Fig. 45 Scattering patterns from the PEP-PE stars at $\Phi_{\mathrm{pol}} = 0.24\%$ for three different temperatures. The lower temperatures data exhibit a Q^{-1} power law behavior at low Q

at room temperature. The rods must be longer than ~100 nm such that only the asymptotic law is observed within the Q range available. Evaluation of the structural and density parameters following Eq. 62 and knowing that all polymers are aggregating leads to a rod radius of $a = 146 \pm 1$ Å, and the overall polymer volume fraction within the rod $\Phi_{\mathrm{pol}}^{\mathrm{agg}} = 0.24$.

It was concluded that the placement of the crystalline blocks either at the rim or in the center of the corona controls the identity of self-assembled structures. The PE blocks at the corona rim are little restricted and crystallize in their normal plate-like structure dictated by the crystallization enthalpy. When they are at the star center the outside amorphous corona strongly shields the crystallization event from interaction with other stars. Thus, in the latter case, the corona screens to a large extent the cocrystallization events

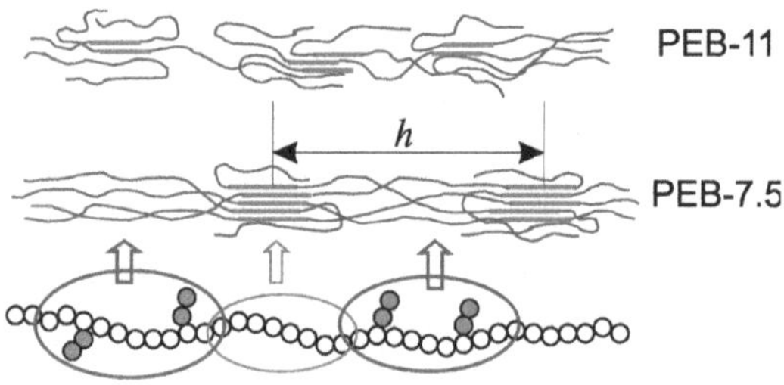

Fig. 46 Cartoon of the structures formed by the PEB-n random copolymers in solution. The amorphous segments containing ethyl side groups screen the cocrystallization between uninterrupted methylene sections of different chains. The rare cocrystallization events lead to formation of one-dimensional aggregates. In the case of PEB-7.5 (a copolymer presenting longer uninterrupted methylene sections than PEB-11) such cocrystallization effects lead to occurrence of density modulation along the rod-like aggregates as revealed by SANS measurements

between different stars. The few rare cases where inter-star crystallization occurs leads to one-dimensional growth, assembling stars such as pearls on a necklace. It was suggested that the same mechanism also governs the surprising rod formation of the PEB-*n* random copolymers where cocrystallization events between different chains are screened and one-dimensional growth is promoted. A cartoon of such a suggested structure is displayed in Fig. 46. In the case of PEB-11, a copolymer with a relatively high density of ethyl side groups and, consequently, a lower crystallization tendency, the cocrystallization of crystallizable sections from different chains is rare. In the case of PEB-7.5, a copolymer containing longer uninterrupted methylene sections, such cocrystallization effects occur more frequently and lead to modulation of crystalline regions along the one-dimensional formed structures.

5.3 Cocrystallization of C_{24} Wax and PEB-11 Random Copolymer

From analysis of the characteristic scattering patterns of PEB-7.5 and PEB-11 random copolymers it was concluded that, although both polymers reveal the same overall aggregation behavior in the low temperature range (a characteristic Q^{-1} power law indicative of rod-like aggregates), the self-assembly commences at very different temperatures ($\leq 0\,°C$ for PEB-11 and between 40 and 60 °C in the case of PEB-7.5). The similarities point to a general tendency of amorphous-crystalline random copolymers to self-aggregate in one-dimensional structures, whereas the differences arise from the mean ethylene segment lengths, which influence the degree of crystallinity. This behavior led to the conjecture that interaction of PEB-*n* copolymers with waxes may depend on the correlation between the polymer self-assembling temperature and the wax solubility point within the investigated range of wax concentration. The density of the ethyl side groups and the M_w may thus control the efficacy of different copolymers in modifying the size and shape of waxy crystals formed in decane. Therefore, in order to elucidate the interaction mechanism, different polymer–wax systems have been the subject of systematic SANS investigations performed within a very wide temperature range (from 85 to −30 °C) and under different contrast conditions. This allows the visualization of either the wax or the copolymer and, in some cases, both components become visible.

The solubility line for the series of waxes in *d*-decane was determined by observation of the CP [12]. Figure 47 shows that the wax solubility decreases with increasing carbon number. Below the solubility lines the waxes precipitate in large compact plate-like structures nicely revealed by optical microscopy (Fig. 48). Such crystals provide a strong scattering intensity in comparison with that of the single wax molecule form factor. Figure 49 reveals the scattering cross-sections from different amounts of $C_{24}H_{50}$ wax in *d*-decane measured at −20 °C, whereas Fig. 50 presents the results obtained

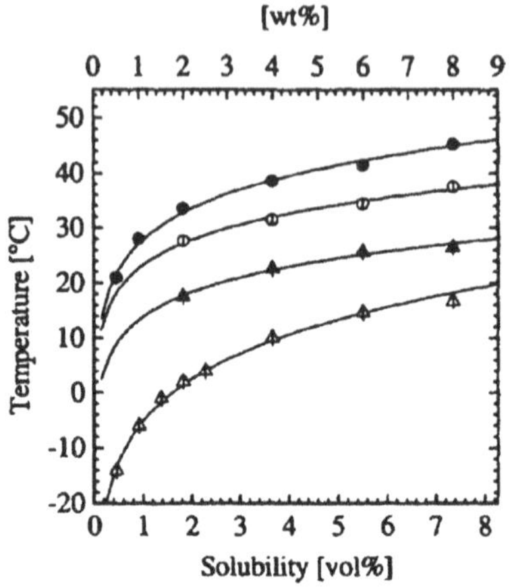

Fig. 47 Observed CP of waxes in decane: C24H50 (*open triangles*), C28H58 (*closed triangles*), C32H66 (*open circles*), and C36H74 (*closed circles*). The solubility lines are obtained by fitting the data with the van't Hoff equation [16]

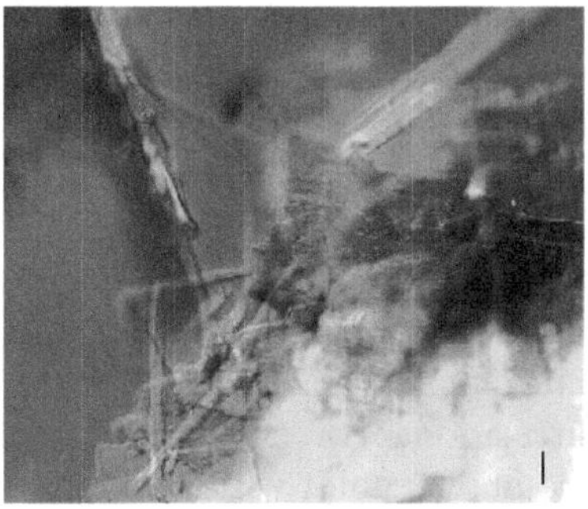

Fig. 48 Micrograph of a 2% C_{36} wax solution in decane at room temperature; the *scale bar* represents 20 μm

from different amounts of $C_{36}H_{74}$ wax in *d*-decane at 10 °C. As we can observe, the scattering patterns differ significantly although crystals formed by both waxes give rise to a strong increase of forward scattering. For the shorter wax over a wide Q range a Q^{-3} power law is observed. In addition, a fraction of the wax molecules remain in solution.

This may be deduced from the scattering pattern at high-Q, which displays the wax molecule form factor. The power law indicates formation of three-dimensional fractal-like structures without a well-defined surface, i.e., open structures made by agglomerating aggregates like the house-of-cards morphology (Fig. 2). The scattering profiles characteristic for the longer wax agree with Porod scattering, indicating the formation of large compact crystals such as those revealed by the micrographs (Fig. 48). Such objects can

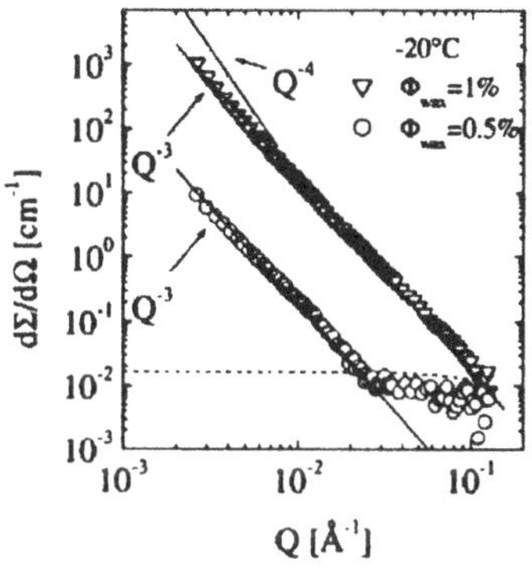

Fig. 49 SANS profiles from C_{24} solutions in *d*-decane at –20 °C for 0.5% (*open circles*) and 1% (*open triangles*) wax contents. The *solid lines* depict the power law behavior of the scattered intensity, while the *dotted line* represents the single wax molecule form factor in *d*-decane for a 0.5% wax volume fraction

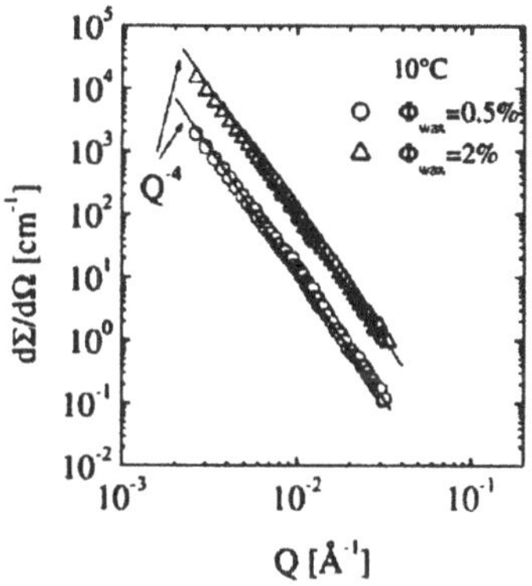

Fig. 50 SANS profiles from C_{36} solutions in *d*-decane at 10 °C for 0.5% (*open circles*) and 2% (*open triangles*) wax contents. The *lines* have the same meaning as in Fig. 49

form superstructures that at a larger scale may exhibit the house-of-cards morphology, reaching out hundreds of micrometers and ultimately leading to gelling.

Comparing the wax solubility lines (Fig. 47) with the temperature evolution of the scattered intensity from PEB-*n* random copolymers in *d*-decane (Fig. 37) we observe that a good match of the wax and polymer aggregation tendency occurs if, e.g., small amounts of C_{24} wax ($\leq$2%) are mixed with the PEB-11 copolymer, or if larger amounts of C_{36} wax are mixed with PEB-7.5 copolymer. In these cases a cocrystallization of wax and polymer in common structures could possibly take place. For this reason detailed SANS studies of such wax–polymer mixtures for different contrast conditions (according to Eq. 11) were done in order to elucidate the interaction mechanism taking

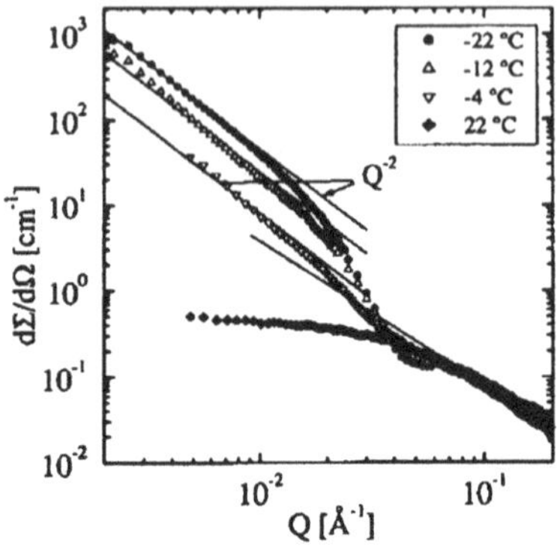

Fig. 51 Temperature dependence of the SANS patterns from a mixed solution of PEB-11 ($\Phi_{\text{pol}} = 1\%$) and C_{24} wax ($\Phi_{\text{wax}} = 1\%$) in decane under polymer contrast. The *solid lines* indicate the Q^{-2} power law behavior at low Q

place and the evolving structures. Figure 51 presents temperature-dependent SANS patterns from a mixed solution of 1% PEB-11 and 1% C_{24} wax, where the wax was matched by the solvent and thus only the copolymer was visible. The difference to Fig. 36 (top), where the self-assembly of the polymer alone is displayed, is obvious: the addition of wax alters the polymer self-assembly completely, now yielding two-dimensional aggregated structures. This is demonstrated by the data at lower temperatures where $\mathrm{d}\Sigma/\mathrm{d}\Omega \sim Q^{-2}$ characteristic for plate scattering, is observed.

Thus, the presence of the wax changes the aggregation behavior of PEB-11 from rod-like to plate-like structure formation. These plate-like aggregates commence to appear first at $-4\,°\mathrm{C}$, while at $22\,°\mathrm{C}$ the single chain cross-section is primarily visible. The data indicate that the wax cocrystallizes with the PE segments of the random copolymer. Unlike the platelets formed by the PE-PEP diblock, which act as a nucleation platform for waxes and thus does not change their morphology, the self-assembling structure of PEB-11 is altered by the presence of wax. In the high-Q regime the intensity drop and the well-defined minimum exhibited around $Q = 0.04\,\text{Å}^{-1}$ relate to the form factor for a finite platelet thickness. An evaluation in terms of the two-dimensional Guinier approximation (Fig. 52) according to Eq. 68 yielded the parameters reported in Table 9.

Figure 53 compares the experimental scattering profile at $-22\,°\mathrm{C}$ with the evaluated form factor for an infinite plate with a perpendicular rectangular density profile $C_{\mathrm{P}}(z)$ of width d_{eff} derived from Eq. 25 and Eq. 30:

$$P(Q) = \left| \int\limits_{-\infty}^{+\infty} C_{\mathrm{P}}(z) \exp(\mathrm{i}Qz)\,\mathrm{d}z \right|^2 = \left(\sin(Qd_{\text{eff}}/2)/(Qd_{\text{eff}}/2)\right)^2 . \tag{69}$$

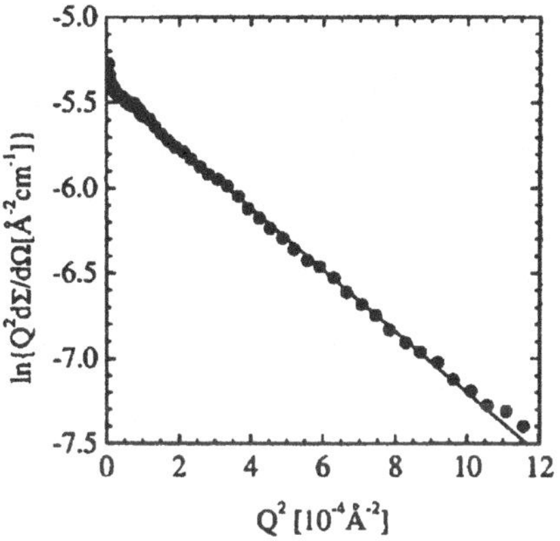

Fig. 52 SANS result from a solution of PEB-11 ($\Phi_{pol} = 1\%$) and C_{24} wax ($\Phi_{wax} = 0.5\%$) at -22 °C in decane under polymer contrast in a Guinier presentation

Table 9 Polymer and wax aggregation characteristics for the PEB-11/C_{24} mixed system in decane ($\Phi_{pol} = 1\%$)

Φ_{wax}	Temp. °C	Polymer d_{eff} Å	Polymer $Q^2(d\Sigma/d\Omega)$ 10^{13} cm^{-3}	Wax $Q^2(d\Sigma/d\Omega)$ 10^{13} cm^{-3}	Wax[a] P_4 10^{25} cm^{-5}
0.005	−4	–	0.4 ± 0.05	0.03 ± 0.01	–
	−12	121 ± 2.3	1.89 ± 0.05	0.11 ± 0.015	–
	−22	147 ± 0.4	4.52 ± 0.02	0.68 ± 0.007	–
0.01	−4	88 ± 1	0.8 ± 0.1	0.17 ± 0.03	–
	−12	133 ± 2	2.2 ± 0.06	0.5 ± 0.04	–
	−22	149 ± 1	4.8 ± 0.03	2.12 ± 0.02	–
0.2	−4	87 ± 1	0.76 ± 0.05	0.52 ± 0.05	1.9
	−12	100 ± 0.5	0.99 ± 0.04	0.75 ± 0.07	1.9
	−22	149 ± 0.4	4.1 ± 0.05	1.1 ± 0.1	3.6

[a] Porod constant from the Q^{-4} part in the cross-section

The fitted form factor (dashed line) containing, in addition to Eq. 69, the asymptotic diffuse scattering from the monomer correlations within the chains (solid line) yields a thickness $d_{eff} = 132$ Å, agreeing favorably with that from the Guinier approximation (Table 9).

Figure 54 presents the temperature-dependent scattering profiles from the same polymer–wax combination under wax contrast. The data displays a very pronounced Q^{-2} behavior up to high Q, leading to the conclusion that the wax platelets need to be very thin since no two-dimensional Guinier regime is visible. The extended Q^{-2} pattern is compatible with plates from a single layer of stretched C_{24} molecules ($d_{eff} = 32$ Å) and excludes formation of stacks of

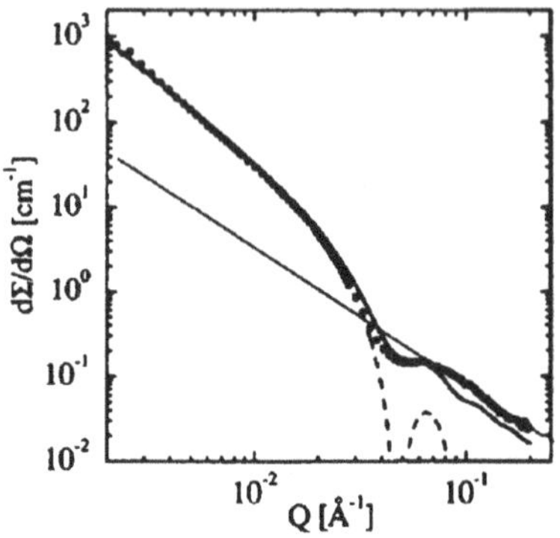

Fig. 53 Model description of the scattering pattern under polymer contrast from a mixed decane solution of 1% PEB-11 and 1% C_{24} at −22 °C according to Eq. 69

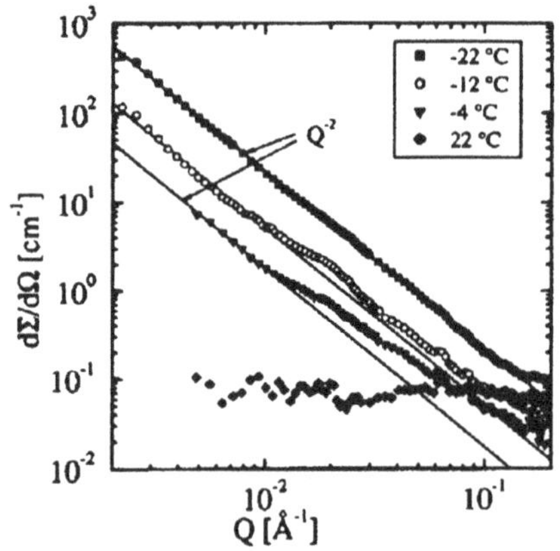

Fig. 54 Temperature dependence of the SANS patterns from a mixed solution of PEB-11 (Φ_{pol} = 1%) and C_{24} wax (Φ_{wax} = 1%) in decane under wax contrast. The *solid lines* indicate the Q^{-2} power law behavior of the scattered intensity observed over a wide Q range

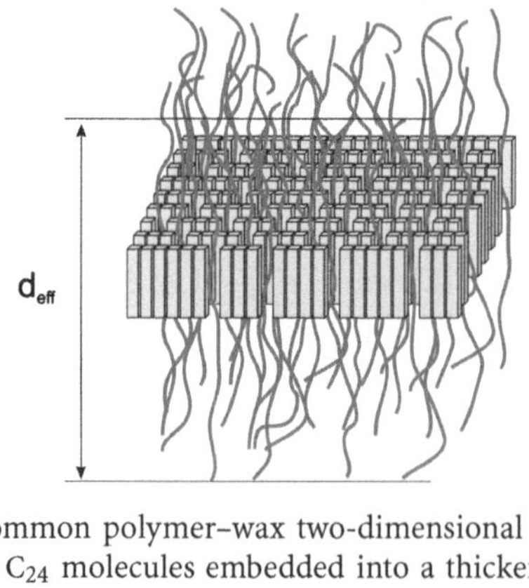

Fig. 55 Cartoon of the common polymer–wax two-dimensional aggregates consisting of a monolayer of stretched C_{24} molecules embedded into a thicker homogeneous polymer platelet of thickness d_{eff}

more molecular layers. It is significant to observe that already at –4 °C, where according to the phase diagram [12] all wax should be in solution, a pronounced increase in the low-Q scattering is observed.

The experimental findings suggest the cocrystallization of wax with polymer. This process yields platelet structures like those cartooned in Fig. 55 where the aggregates involve a crystalline thin layer of wax and cocrystallized polymer sections. Using the contrast-matching SANS results the asymptotic forward scattering from Eq. 68 may be written for each of these two embedded layers:

$$\left.\frac{d\Sigma}{d\Omega}Q^2\right|_{\text{wax}} = \Phi_{\text{wax}}^{\text{plate}}\,\Delta\rho_{\text{wax}}^2\left(\Phi_{\text{wax}}^{\text{layer}}\right)^2 2\pi d_{\text{wax}}\,, \tag{70}$$

$$\left.\frac{d\Sigma}{d\Omega}Q^2\right|_{\text{pol}} = \Phi_{\text{pol}}^{\text{plate}}\,\Delta\rho_{\text{pol}}^2\left(\Phi_{\text{pol}}^{\text{layer}}\right)^2 2\pi d_{\text{pol}}\,,$$

where $\Phi_{\text{wax}}^{\text{plate}}$ and $\Phi_{\text{pol}}^{\text{plate}}$ are the respective volume fractions of wax and polymer plates in solution, $\Delta\rho_{\text{wax}}$ and $\Delta\rho_{\text{pol}}$ the wax and polymer scattering contrast, and $\Phi_{\text{wax}}^{\text{layer}}$ and $\Phi_{\text{pol}}^{\text{layer}}$ the wax and polymer concentrations within the layers. The volume fractions and concentrations of wax and polymer in different states may be related to each other and to geometrical parameters using the relations presented in the following:

(i) The volume fraction of plates is related to the volume fraction of polymer and wax by $\Phi^{\text{plate}} = (\Phi - \Phi^{\text{sol}})/\Phi^{\text{layer}}$, where Φ^{sol} is the fraction of polymer or wax still in solution

(ii) The homogeneous polymer distribution across the layer experimentally observed leads to $\Phi_{\text{pol}}^{\text{layer}} = (1 - \Phi_{\text{wax}}^{\text{layer}})$

(iii) A joint polymer and wax platelet structure means that $\Phi_{\text{pol}}^{\text{plate}} = \Phi_{\text{wax}}^{\text{plate}}$ $(d_{\text{pol}}/d_{\text{wax}})$

Using Eq. 70 with these relations implemented, the measured thickness and forward scattering (Table 9), the known contrast factors (Table 4), the volume fraction of polymer and wax within the layers, as well as the total aggregated volume fractions have been estimated. These are reported together with the platelet area per cubic centimeter, $A_{\text{wax}}/V = \Phi_{\text{wax}}^{\text{plate}}/d_{\text{wax}}$ in Table 10.

For a larger wax content (2%) a strong enhancement of the intensity with a Q^{-4} behavior was observed at low-Q within the wax contrast scattering patterns [11]. This scattering profile suggests the presence of larger three-dimensional wax crystals coexisting with plate-like structures. The scattering under polymer contrast from the same sample composition always relates to two-dimensional structures. Following the interpretation of the evaluated Porod constant Eq. 22 and supposing crystals of spherical shape, the estimated average size of the three-dimensional wax objects was $\cong 2400$ Å. From a qualitative and quantitative interpretation of the SANS results in the same

Table 10 Density parameters of the mixed PEB-11/C_{24} platelets formed in decane ($\Phi_{pol} = 1\%$)

Φ_{wax}	Temp. °C	Φ_{wax}^{layer}	$\Phi_{wax}^{plate}\Phi_{wax}^{layer}$	Φ_{pol}^{layer}	$\Phi_{pol}^{plate}\Phi_{pol}^{layer}$	Area plate $10^4\ cm^2/cm^3$	Area/Φ_{wax} $10^6\ cm^2/cm^3$
0.005	– 4	0.55	0.00017	0.45	0.0004	0.10	0.16
	– 12	0.62	0.00053	0.38	0.0014	0.31	0.56
	– 22	0.76	0.0028	0.24	0.0043	1.22	2.25
0.01	– 4	0.69	0.0008	0.31	0.001	0.37	0.27
	– 12	0.78	0.002	0.22	0.0025	0.85	0.77
	– 22	0.87	0.0077	0.153	0.0067	2.94	2.75
0.2[a]	– 4	0.8	0.002	0.2	0.0015	0.86	0.41
	– 12	0.83	0.00284	0.17	0.00194	1.14	0.56
	– 22	0.81	0.0043	0.19	0.005	1.77	0.78

[a] Platelet fraction only

manner as for the lower wax content it was concluded [11] that a staged mechanism where polymer and wax form two-dimensional aggregates (which then nucleate three-dimensional wax crystals) is very probable.

The experimental results showed that the polymer and wax influence each other's aggregation behavior. The presence of C_{24} wax changes the aggregation behavior of PEB-11 from rod to platelet structure formation. Two-dimensional monolayer crystals of C_{24} wax embedded into thick homogeneous polymer layers were observed instead of agglomerating wax crystals forming house-of-cards morphologies. A cocrystallization process, where wax and polymer jointly form platelet structures, was revealed. It seems that the conditions for the formation of these aggregates are dictated by a very delicate balance between the enthalpy gain from the cocrystallization and the entropy loss from the conformation entropy of the polymer. The occurrence of the cocrystalline state instead of self-aggregation indicates a preferential free energy situation, which could come from a relatively smaller entropy loss due to the paraffin process in the polymer-crystalline aggregates.

5.4 Templating and Cocrystallization of Waxes and PEB-7.5 Random Copolymers

The experimental findings for mixed solutions of PEB-11 random copolymer and C_{24} wax confirmed the idea that a good match of the polymer self-assembling temperature with the wax CP promotes a mechanism of cocrystallization of wax and polymer within common aggregates. Such a process keeps the wax crystals at moderate sizes and arrests the growth of large three-dimensional objects forming house-of-cards arrangements. This observation stimulated the investigation of another wax–copolymer combination fulfilling the matching condition, namely that of PEB-7.5 (6 K) random copolymer with larger amounts of C_{36} wax (~8%). In this case, the self-association of both components should take place above 40 °C, as indicated by the wax solubility line and polymer SANS results.

Nevertheless, unexpected SANS results have been obtained from $C_{36}H_{74}$ solutions in *d*-decane at high temperatures. Figure 56 shows the volume fraction normalized scattering profiles from 8% and 2% wax solutions at 80 °C, well above the solubility line of C_{36} wax. The pronounced intensity increase observed at low-Q follows a Q^{-2} power law. It appears that even well above the solubility a certain fraction of wax molecules forms plate-like structures that coexist with the free wax molecules in solution. The relative amount of the aggregates seems to be independent of the wax volume fraction. This observation led to the conclusion that in combination with the PEB-7.5 copolymer, the C_{36} tendency to form platelets at higher temperatures will drive the wax–copolymer interaction mechanism and control the morphology independently of the wax concentration in solution. Figure 57 displays the scattering profiles at 77 °C from the interacting system of C_{36} wax and PEB-7.5

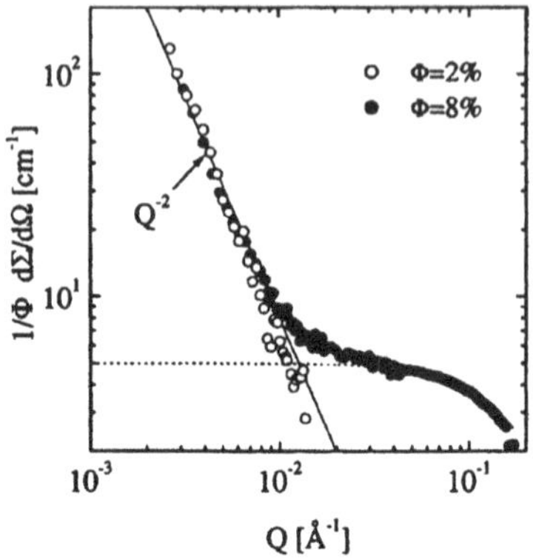

Fig. 56 Volume fraction normalized SANS patterns from two solutions of C_{36} wax in *d*-decane at 80 °C. The *solid line* indicates the asymptotic power-law regime observed at low Q, while the *dotted line* represents the single molecule form factor

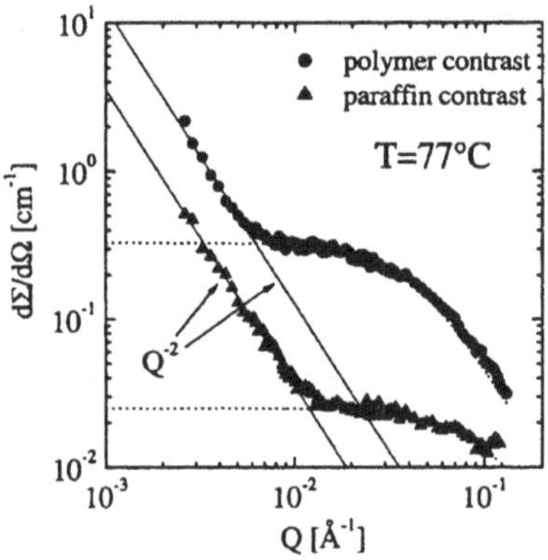

Fig. 57 SANS patterns obtained at 77 °C from a mixed solution of 6 K PEB-7.5 (Φ_{pol} = 0.6%) and C_{36} (Φ_{wax} = 0.5%) in decane under polymer (*full circles*) and wax (*full triangles*) contrast. The *lines* have the same meaning as in Fig. 56

copolymer in *d*-decane for 0.5% wax and 0.6% polymer volume fractions. The polymer and wax were made visible by contrast matching according to Eq. 11. The two-dimensional aggregates formed by the wax molecules were observed even at this low wax content, where a clear Q^{-2} power law behavior of the scattered intensity is revealed by the wax profile toward low-*Q*. The corresponding polymer profile also displays such a low-*Q* behavior indicating the formation of a similar morphology by the polymer. This situation is in contrast to the pure PEB-7.5 behavior in *d*-decane, where the polymer stays in the single coil conformation at this temperature. Apparently, the wax mediates the formation of common wax–polymer plates.

At lower temperatures, a massive joint aggregation process of wax and copolymer, similar to that revealed by the PEB-11/C_{24} combination, was ob-

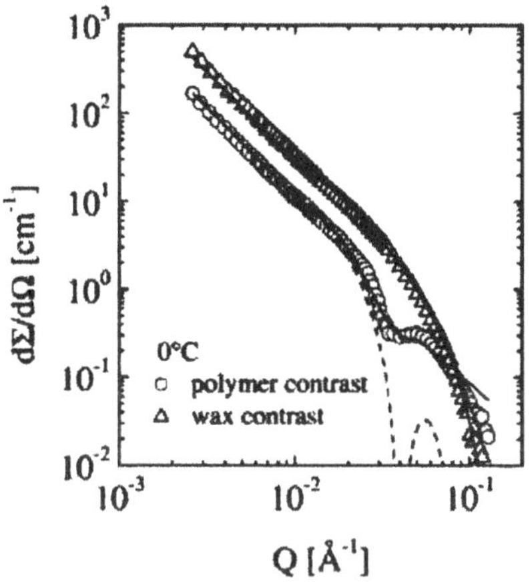

Fig. 58 Polymer and wax scattering patterns at 0 °C from the same solution as in Fig. 57. The *solid lines* represent the model description of the data according to the text (combined Eqs. 29–31 and 38) while the *dashed line* represents the polymer platelet

served. Figure 58 shows the scattering profiles from the same PEB-7.5/C_{36} system at 0 °C. For the wax case, the Q^{-2} power law extends over a wide Q range and the high-Q deviation from the asymptotic behavior yields information on the thickness of the wax plates (about 40 Å, again a wax monolayer). For the polymer contrast, the scattering pattern behaves like Q^{-2} only at low-Q, while towards higher Q values it transforms into an oscillatory behavior around the power law, showing a well-defined minimum followed by a maximum; a profile similar to the PEB-11 scattering pattern in the presence of C_{24} wax. On the basis of this qualitative observation, it was concluded that the cocrystallization process leads to the formation of thin wax layers embedded into thicker amorphous polymer plates displaying a homogeneous polymer distribution. The scattering patterns were commonly fitted using the above-defined model by Eq. 70 and its following definitions. As in the case of the PEB-11/C_{24} combination for higher wax contents at lower temperatures, a strong low-Q upturn of the intensity reveals formation of large three-dimensional aggregates which, besides wax, this time involve certain amounts of polymer. Obviously, the wax surplus grows into the third dimension from the polymer–wax platelets. The structural and density parameters obtained from the fit of all C_{36}/PEB-7.5 combinations, following a similar approach as for the C_{24}/PEB-11 case, are reported in Table 11.

The influence of the PEB-*n* copolymers on the size and shape of the wax crystals formed in solution may be observed in Fig. 59, where micrographs of a 2% C_{36} solution in *d*-decane including 1% PEB-7.5 are presented with different magnification scales. The effect of polymer addition is striking: instead of large sharp waxy crystals (Fig. 48) much smaller and softer morphologies are formed.

Table 11 Structural and density parameters of the mixed PEB-17.5/C_{36} platelets formed in decane ($\Phi_{pol} = 0.6\%$)

Φ_{wax}	Temp. °C	d_{eff} polymer Å	Φ_{wax}^{layer}	$\Phi_{wax}^{plate}\Phi_{wax}^{layer}$ $(1/\Phi_{wax})$	Φ_{pol}^{layer}	$\Phi_{pol}^{plate}\Phi_{pol}^{layer}$ $(1/\Phi_{pol})$	Plate area 10^4 cm^2/cm^3	P_4 (wax) 10^{-7} cm^{-1} Å^{-4}
0.005	20	125	0.77	0.38	0.23	0.27	0.54	–
	0	135	0.81	0.80	0.19	0.484	1.1	–
0.02	30	170	0.89	0.31	0.11	0.45	1.53	–
	0	185	0.91	0.55	0.09	0.68	2.66	4.5
0.4	40	160	0.79	0.03	0.21	0.22	0.4	1.8
	0	175	0.88	0.15	0.12	0.48	1.44	9

$\Phi_{wax}^{plate}\Phi_{wax}^{layer}$ and $\Phi_{pol}^{plate}\Phi_{pol}^{layer}$ represent the volume fraction of the aggregated wax and copolymer (plates only)

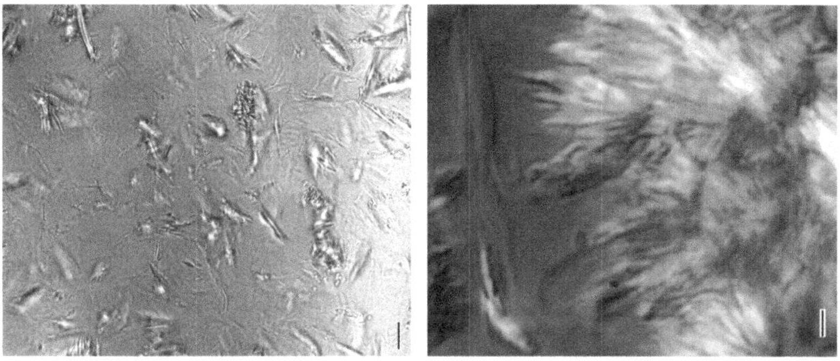

Fig. 59 Micrographs at room temperature of PEB-7.5(6 K)-treated C_{36} solution in decane ($\Phi_{pol} = 1\%$, $\Phi_{wax} = 2\%$); *scale bars* represent 20 μm (*left*) and 2 μm (*right*)

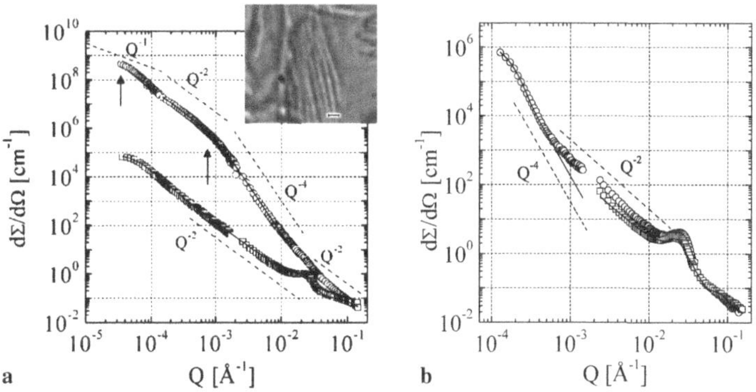

Fig. 60 Scattering patterns under **a** wax and **b** polymer contrast from a mixed solution of 0.6% 6 K PEB-7.5 copolymer and 4% C_{24} wax in *d*-decane at 0 °C (*open squares*) and -10 °C (*open circles*). The *dashed lines* show the power laws observed within different *Q* ranges, while *full lines* represent the fit of the data according to the models described in the text. The characteristic sizes of the two different types of wax aggregates identified at -10 °C are marked by *arrows*. The *insert* shows a micrograph of the aggregate formed under the same conditions as in the scattering experiments. The *scale bar* denotes 10 μm

Now we address the situation where the PEB-*n* copolymers self-assemble at temperatures higher than the wax solubility line. Intuitively, in this situation we would expect the wax–copolymer aggregation to be controlled by the polymer self-assembling behavior. Figures 60a and b present separately the polymer and wax scattering patterns from a solution of 0.6% low M_w PEB-7.5 (6 K) copolymer and 4% C_{24} wax in *d*-decane at two temperatures within the common aggregation regime (well below the temperature at which polymer rods appear, between 40 and 20 °C; Fig. 36 bottom). The scattering patterns were collected over a wide *Q* range by combining classical SANS, FSANS, and USANS measurements. Obviously, multilevel structures displaying characteristic length scales from 1 nm to 10 μm are formed upon cooling.

From analysis of the scattering profiles within the classical SANS Q range (above 0.002 Å^{-1}) at 0 °C under polymer and wax contrast, it was concluded that the polymer and wax influence each other's aggregation behavior. On the one hand, the wax aggregation influences the polymer structure, which changes from rod-like (Q^{-1} power law) in the case of a pure polymer solution to plate-like (Q^{-2} power law) in the presence of wax. Thus, a cocrystallization of the polymer and wax occurs again below certain temperatures. On the other hand, both the wax and polymer scattering profiles reveal a correlation peak at the same Q value, where it is also found for the polymer alone (around $Q^* = 0.025$ Å^{-1}). Thus, the primordial one-dimensional density-modulated polymer structure formed at higher temperature influences significantly the joint aggregation mechanism and leads to the occurrence of correlated wax–copolymer platelets. USANS and FSANS data under wax contrast indicated very large platelets as the Q^{-2} power law behavior of the scattered intensity extends over a very wide Q range. By USANS, a Guinier regime was observed that indicates a lateral extension of platelets of several micrometers. Such a morphology, consisting of platelets forming correlated arrangements, is reminiscent of the platelet staples observed in the case of high M_W PE-PEP diblocks.

Similarly, the correlation effects were described in terms of a paracrystalline structure factor Eq. 41. The common wax copolymer structures were characterized in terms of an ensemble of finite-sized disks, each disk consisting of a wax- and polymer-containing lamellar similar to the description of the PEB-11/C_{24} or PEB-7.5/C_{36} two-dimensional aggregates Eq. 70. A common fit at both contrast conditions in terms of this model [14] provides a good description of the scattering profiles at 0 °C and reliable structural and density parameters of the wax–copolymer aggregates (see Table 12).

At −10 °C the scattering profiles of both components differ from one another:

(i) The polymer still retains the two-dimensional correlated structures as demonstrated by the Q^{-2} power law and the correlation peak. The low-Q data measured by FSANS reveal the scattering from an additional structure, the large compact polymer crystals formed at much higher temperatures. The features of this structure were analyzed in terms of the Beaucage model (Eq. 20) with a power law exponent ($P = 4$). Comparison with the case of the pure polymer self-assembly shows that the addition of wax has only a minor effect on the features of these aggregates.

(ii) The wax scattering pattern at −10 °C revealed that the wax massively crystallizes and grows from thin platelets into large objects with sharp interfaces (Q^{-4} power law at high-Q). Within the Q range covered by FSANS ($Q < 10^{-3}$ Å^{-1}) Q^{-2} power laws were found, indicating that these objects are thick plates. In terms of the unified Beaucage equation for multiple structural levels [100, 101], a thickness of 0.23 μm was identified. From the forward scattering it was concluded that at −10 °C 18% of the wax

Table 12 Parameters of the common PEB-7.5/wax aggregates templated by primordial density-modulated polymer rods in decane obtained from the fit of the data measured over the entire Q range available by USANS, FSANS and classical SANS; fit with the correlated finite platelets model

M_W/Φ_{pol}	Wax/Φ_{wax}	Temp. °C	d_{eff} polymer Å	d_{eff} wax Å	$D-\sigma_D$ Å	R µm	$\Phi_{pol}^{plate}\Phi_{pol}^{layer}$ $(1/\Phi_{pol})$	$\Phi_{wax}^{plate}\Phi_{wax}^{layer}$ $(1/\Phi_{wax})$
6 K/0.006	C_{24}/0.04	0	110	32	223–101	3.2	0.33	0.04
	C_{24}/0.04	−10	–	2300	–	3.2	–	0.18
6 K/0.006	C_{24}/0.005	−20	81	32	191–136	3.1	0.45	0.10
30 K/0.01	C_{36}/0.04	20	155	45	503–327	–	0.35	0.20

$\Phi_{wax}^{plate}\Phi_{wax}^{layer}$ and $\Phi_{pol}^{plate}\Phi_{pol}^{layer}$ represent the volume fraction of the aggregated wax and copolymer (thin or thicker plates); R represents the radius of the common polymer–wax disk-like platelets

in solution was combined in these aggregates. At even lower Q some additional scattering was visible, indicating the formation of even larger aggregates. Their characterization was difficult due to the lack of sufficient data points in the Guinier range.

Tentatively, a one-dimensional morphology was attributed to the largest aggregates in assigning a Q^{-1} power law to these low-Q data. This assignment was confirmed by optical microscopy observations of large rods at −10 °C (inset of Fig. 60). The observations summarized above were interpreted in terms of a hieratical structure formation. The primordial structure is created by the self-assembly of the polymers into rod-like aggregates with a longitudinal alteration of crystalline and amorphous regions. Then, wax crystallization commences at a lower temperature. It is apparently mediated by the crystalline parts of the modulated polymer structure. The wax structures grow laterally involving additional polymer chains still in solution, as schematically shown in Fig. 61. Thus, large common wax–polymer platelets develop around the rods as a secondary morphology. This explains the change of scattering behavior from Q^{-1} in the case of wax-free polymer self-assembling, to Q^{-2} in the case of wax–polymer solutions.

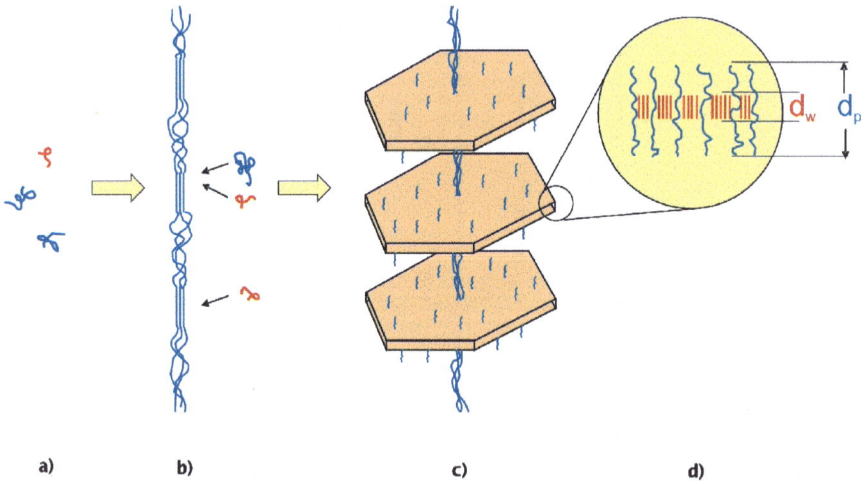

Fig. 61 Cartoon presenting the mechanism of wax crystallization on polymer templates. According to the experimental findings such a mechanism yields multiscale structures evolving hierarchically with decrease in temperature: **a** the single coil stage; **b** the formation of polymer rods with modulated density (alternation of crystalline and amorphous sequences along the rod axis); **c** the shish-kebab-like morphology formed by correlated polymer–wax plates templated by the initial polymer rods; **d** a detail of the polymer–wax common plates, basically the same structure as that shown in Fig. 55. At lower temperatures and high wax content the wax surplus fills the space between the platelets and transforms the shish-kebab into large-scale one-dimensional object

Thereby, the platelets contain most of the material while the primordial rods serve to template the lateral discs in a correlated arrangement resembling a shish-kebab, similar to the morphologies formed by crystallization-induced flow orientation [102, 103]. The main requirement for the occurrence of such morphology is the generation of a fiber as a template for the later growing of lateral platelets. In the case of polyethylene and cellulose [104], the fibers crystallize during flowing. In the present case, the orientation mechanism is based upon the tendency of the polymer to aggregate into a graded one-dimensional structure prior to the inclusion of the paraffin.

Later crystallization stages lead to a thickening of the platelets by the wax surplus, which grows from the secondary two-dimensional structures. In this way the correlation effect disappears from the wax scattering pattern and the thicker platelets observed at −10 °C are formed. These platelets become thicker and join, leading to the appearance of a compact one-dimensional tertiary morphology in which the wax encapsulates the primordial rod-like polymer aggregate, as revealed by optical microscopy.

Thus, SANS and microscopy observations, in a complementarily fashion, revealed that the aggregates evolve one from another with decreasing temperature and form a hierarchical morphology having multiple sized structures. The micrographs also provided an indirect proof of the thin polymeric rods. The crystallization of C_{24} wax into one-dimensional objects in the presence of PEB-7.5 is driven by the existence of the primordial rod-like polymer aggregates, which dictate the overall morphology. The polymer rods are decorated by the wax and can be indirectly detected via microscopy (Fig. 62).

When the high M_W PEB-7.5 (30 K) copolymer is mixed with C_{36} wax in decane, almost the same morphological and structural features are observed as in the case of low M_W copolymer and C_{24} wax [14]. The only difference is that, according to the elevated precipitation temperature of both components [12, 13], the joint wax–copolymer correlated platelets occur at much

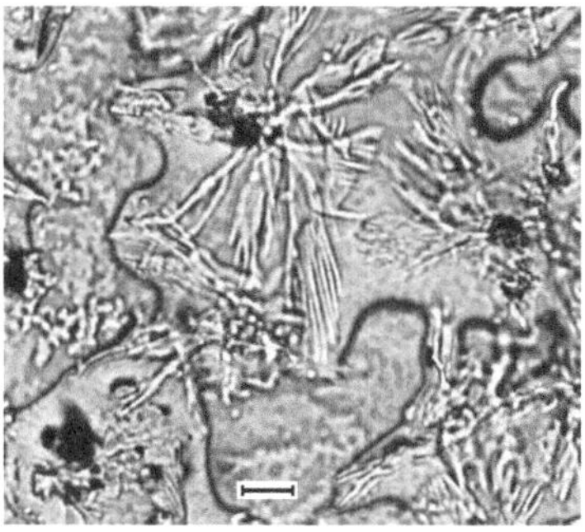

Fig. 62 Optical micrograph from a decane solution of 6 K PEB-7.5 copolymer and 4% C_{24} wax at −10 °C (*scale bar* 30 μm)

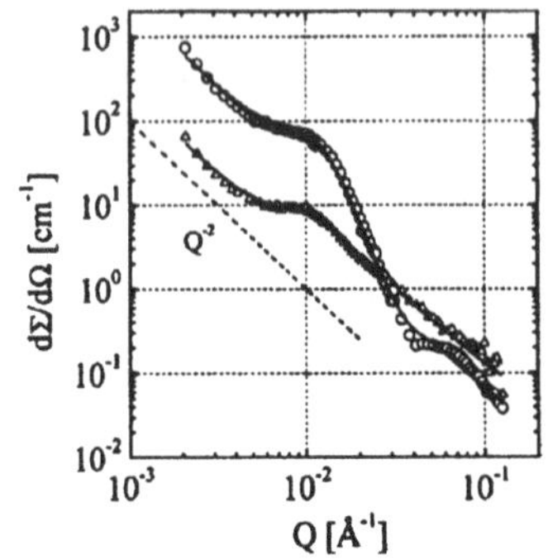

Fig. 63 Scattering patterns under polymer (*open circles*) and wax (*open triangles*) contrast from a mixed decane solution of 1% 30 K PEB-7.5 and 4% C_{36} wax at 40 °C. The *lines* have the same meaning as in Fig. 60

higher temperatures. Figure 63 presents the polymer and wax scattering patterns measured by classical SANS on a solution of 1% high M_w PEB-7.5 copolymer and 4% C_{36} wax at 40 °C. This is within the joint aggregation regime of this wax–copolymer combination. The same behavior as in the case of low Mw PEB-7.5 mixed with C_{24} wax at 0 °C is observed. The Q^{-2} behavior and the correlation peak are again observed at the same Q value as for the pure polymer solution. This points strongly towards a cocrystallization of the polymer and wax in thin correlated platelets mediated by the primordial one-dimensional density-modulated polymer structures. The model of embedded polymer and wax layers forming a correlated arrangement was again successful in describing the experimental data. At 40 °C the joint polymer–wax platelets contain about 35% of all the polymers and 20% of all the wax in solution. This is demonstrated by the TEM observations of the aggregates formed by the high M_w PEB-7.5 copolymer and C_{36} wax as isolated at room temperature from decane solution. Results of these observations are shown in Fig. 64a and b, which displays stacks of platelets arranged one on top of each other. They resemble a top view of a shish-kebab-like morphology, as inferred from the scattering data. The edges of the very large platelets and the central axis from which they have grown irregularly are readily visible.

From the above observations it was concluded that in the case of mixtures of PEB-*n* random copolymers with waxes exhibiting a lower CP than the polymer self-assembling temperature, a polymer-based templating structure is formed first. This structure then acts as the nucleation center for the wax molecules. Compared with the PEB-11/C_{24} and PEB-7.5 (6 K)/C_{36} situations, the wax crystallization on polymeric templates yields large compact aggregates. An inspection of the largest structures formed in the wax copolymer solutions allowed us to conclude that clogging of a 45 μm filter (the standard size for defining the CFPP technical parameter) would not read-

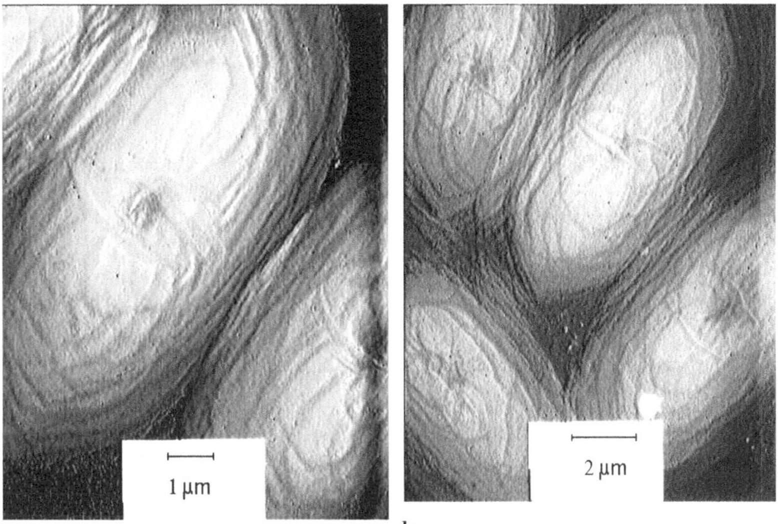

a b

Fig. 64 TEM images of the aggregates isolated at room temperature from a decane solution of of 30 K PEB-7.5 (Φ_{pol} = 1%) and C_{36} wax (Φ_{wax} = 4%). *Scale bars*: 1 µm (**a**) and 2 µm (**b**)

ily occur over a wide range of temperatures in comparison with the undoped wax solutions where compact crystals are formed.

5.5 Yield Stress Studies

Rheological investigations reveal that the PEB-*n* copolymers significantly modify the structure and reduce the yield stress of 4 wt% wax gels [12]. Figure 65 presents the yield stresses of wax gels in decane for four different waxes, C_{24}, C_{28}, C_{32} and C_{36} for a 4% wax volume fraction. We can observe that the yield stresses are almost independent of temperature up to near the solubility boundary where they rapidly diminish. Obviously a sufficient amount of wax must precipitate from solution before a gel network is formed. For the most soluble wax, C_{24}, the gel strength is quite sensitive to temperature. The yield stress of the PEB-*n* doped wax gels have been measured and were normalized by the yield stress of the undoped gels at the same temperature. The C_{28}, C_{32} and C_{36} waxes were studied at 0 °C while C_{24} was investigated at –20 °C, where the yield stress of the undoped wax reached that of the longer waxes. Each wax was mixed with various PEB-*n* copolymers (n = 7.5, 10, 11 and 15). Figures 66 and 67 display yield stress results for the C_{24} and C_{36} waxes. The polymers show qualitatively similar effects on the yielding behavior of these wax gels as the PE-PEP diblocks. In particular, PEB-*n* can reduce the gel yield stress by 3–4 orders of magnitude at polymer

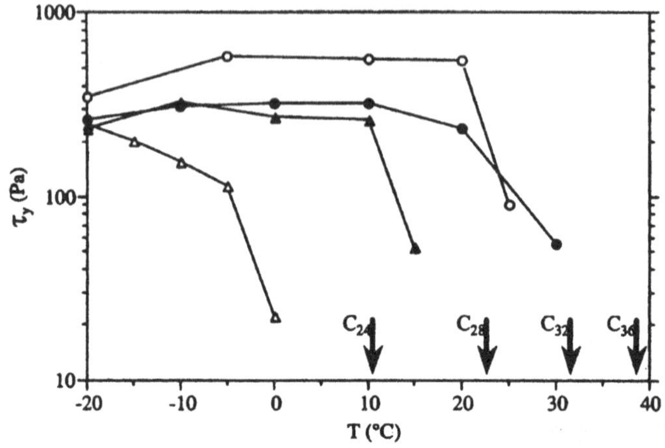

Fig. 65 Measured yield stress of 4% waxes in decane as a function of temperature. The *symbols* denote the C_{24} (*open triangles*), C_{28} (*closed triangles*), C_{32} (*open circles*), and C_{36} (*closed circles*) waxes. The *arrows* indicate the wax CP as extracted from Fig. 46

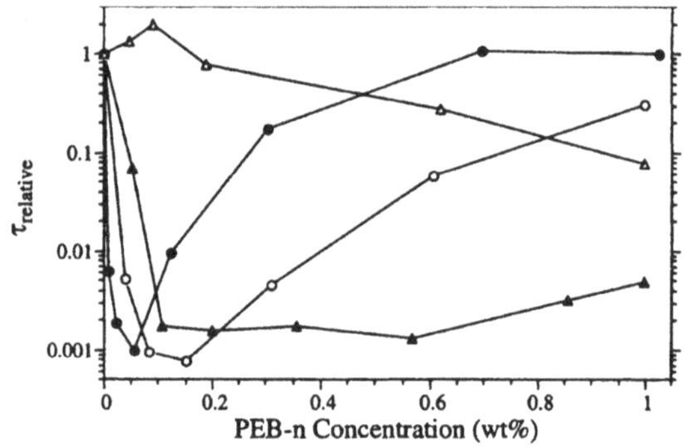

Fig. 66 Effect of PEB-*n* concentration on the yielding behavior of 4% C_{36} wax in decane measured at 0 °C (normalized to the yield stress of undoped wax solution). The *symbols* denote PEB-7.5 (*closed circles*), PEB-10 (*closed circles*), PEB-11 (*closed triangles*) and PEB-15 (*open triangles*) copolymers

concentrations as little as 0.05 wt % or 500 ppm in decane. This is particularly significant for practical applications of this polymer as a wax crystal modifier since it is desirable to use as little of the additive as possible. With increasing PEB-*n* concentration the gel yield stress rebounds and increases in some cases up to levels comparable to the neat gel at polymer concentrations in the order of 1 wt %. Also, in some cases at very low additions of the PEB-*n*, most apparent for PEB-15, polymer additions can increase the gel yield stress.

However, this initial rise is only of the order of a factor of 2 and somewhat overshadowed by the reduction in the yield stress. PEB-*n* copolymers demonstrate selectivity in wax modification depending on the microcrystallinity of their backbone. For the longest C_{36} wax considered, it was found

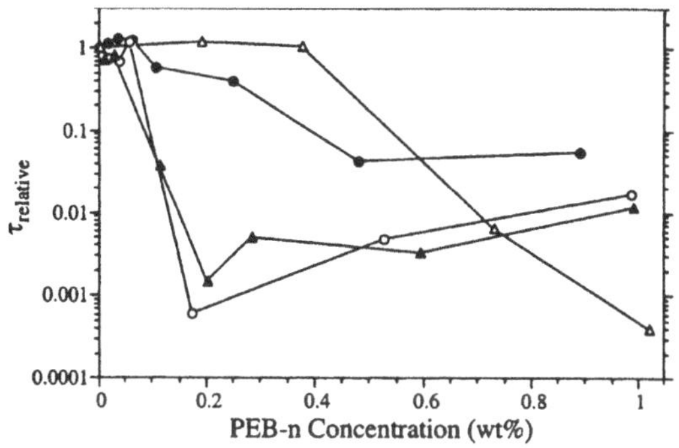

Fig. 67 Effect of PEB-*n* concentration on the yielding behavior of 4% C_{24} wax in decane measured at -20 °C (normalized to the yield stress of undoped wax solution). The *symbols* have the same meaning as in Fig. 66

that PEB-7.5 is the most efficient in reducing the gel yield stress (Fig. 66). It takes less PEB-7.5 to reach the minimum in the wax yield stress than for any of the other PEB-*n*'s examined. The minimum in the gel yield stress is shifted to increasingly higher polymer concentrations with increasing number of ethyl side branches on the polymer backbone, i.e., decreasing polymer microcrystallinity. It is worth noting that the yield stress minimum broadens with an increasing number of ethyl side branches, such that for PEB-11 the gel yield stress is minimal from nearly 0.1 wt % up to 1 wt % although a slight increase in the yield stress is observed at the highest polymer concentrations. PEB-15, in comparison to the other polymers, has almost no effect on the gel yield stress, and a minimum yield stress is not observed below 1 wt % of the polymer in decane. In accordance with the C_{36} wax gels, the polymer concentration monotonically increases from PEB-7.5 to PEB-15 for the C_{32} wax gel [12]. A preference of the polymers towards a certain paraffin length becomes apparent for shorter waxes. For the C_{28} wax gels the yield stress minimum for PEB-7.5 is shifted out to higher concentrations than that for both PEB-10 and PEB-11. This effect is exacerbated for the C_{24} waxes (Fig. 67) where not only is the yield stress minimum for PEB-7.5 shifted to even greater polymer concentrations but in fact PEB-11 may be more efficient at breaking down the wax gel than PEB-10, though this is not a clear differentiation.

The modification of crystal morphologies of C_{36} wax induced by the addition of PEB-7.5 and PEB-10 random copolymers are compared in Fig. 68 as a function of polymer concentration in solution: a clear decrease of the crystal size with increasing the polymer amount is visible, an observation in accordance with the yield stress trend from Figs. 66 and 67. At the highest concentration (0.8%) both morphologies and yield stress effects shown by PEB-7.5 and PEB-10 are similar: numerous crystallites are forming a gel at the same time the increased yield stress observed.

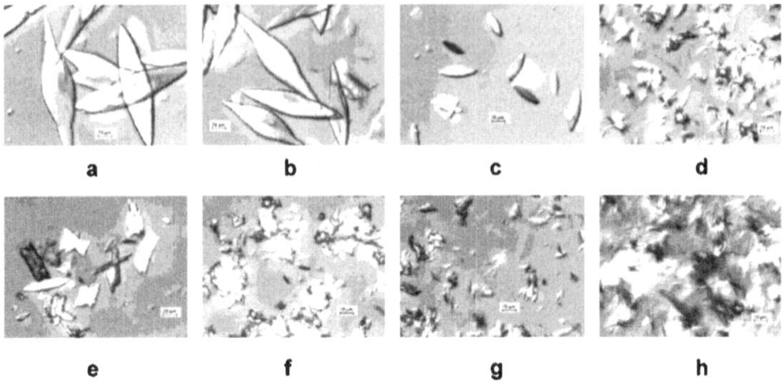

Fig. 68 Optical micrographs of C_{36} crystals formed in a 4% wax solution in decane at 0 °C under the effect of various amounts of PEB-7.5 or PEB-10: **a** 0.05% PEB-7.5, **b** 0.1% PEB-7.5, **c** 0.3% PEB-7.5, **d** 0.8% PEB-7.5, **e** 0.05% PEB-10, **f** 0.1% PEB-10, **g** 0.3% PEB-10, **h** 0.8% PEB-10

We can conclude that the combination of structural investigations by SANS and microscopy with rheological investigations have shown that the poly(ethylene-butene) PEB-*n* random copolymers with variable microcrystallinity displays a high efficiency in controlling the cold flow properties of wax solutions and strongly reduce the size and morphology of wax crystals. The highest efficiency in moderating the wax crystals appears when the polymer and wax cocrystallize in joint structures as a consequence of the good fit between their aggregation properties in solution. The wax crystallization mechanism on polymeric templates is characterized by a lower efficiency as long as produces large scale more compact objects. Nevertheless, the common wax–polymers aggregates formed in this case, though larger, show a size and morphology which hardly can lead to plugging of the filters or pipelines. Finally, the PEB-*n* copolymers show selectivity in their wax modification capacities depending on the ethylene content of the backbone. This suggests that efficient PEB-*n* additives for crude oils and middle distillates should contain graded ethylene contents.

6 Conclusions

The transport and storage of crude oils and their off-springs the middle distillate fuels (diesel, kerosene. and heating oil) is in jeopardy due to the vexatious formation of wax when temperatures approach ~0 °C. Pumping equipment can fail, transmission and fuel lines can plug along with the parallel incapacitation of fuel line filters. Storage tank contents can become immobile and unresponsive to all blandishments for removal and transport. For the kerosene fraction referred to as jet fuel such behavior is laden with catastrophic potential. Clearly these untreated systems (complex fluids) are, to varying degrees, unsatisfactory. While dewaxing may seem to be a facile method to tame these unruly complex fluids, the procedure is expensive and removes the energy-rich high density paraffins.

Historically, diesel fuel has been treated in cold climates with kerosene to, in part, alleviate the fuel line and engine filter plugging problems. Economically and environmentally this treatment mode is no longer acceptable. In an Edisonian style, the copolymers of ethylene and vinyl acetate were found to exert beneficial control (in some fuel systems) over the size of the wax crystals formed. Vinyl acetate-rich materials will yield elastomeric materials whereas the ethylene-rich counterparts are semicrystalline. These copolymers are obtainable in three grades where the vinyl acetate contents range, in mole fraction terms, from 9 to 15%. When successful as a wax crystal additive EVA will produce wax crystals that are smaller and more compact than found in the unmodified systems. Although technologically useful these copolymers, in certain fuels, fail to influence control over wax size leading to a failure in CFPP requirements. A further problem is the precipitation of a fraction of the added EVA prior to the onset of wax formation. Thus, the EVA formulations contain non-active copolymer that fail to contribute to wax crystal control.

Via the combination of the various neutron scattering formats, coupled with optical microscopy, the self-assembling characteristics of copolymers were evaluated. The driving force for aggregate formation was the crystalline segments that alternated with amorphous units. Crystalline-amorphous diblock copolymer-based formulations (PE-PEP materials) have been commercially available (Infinium) since 2000. These materials consist, after self-assembly, of crystalline plates (PE) stabilized in solution by the amorphous (PEP) "hairs". The platelets serve as efficient wax crystal nucleators due to the nucleation properties of the PE plate surface. These plates suppress the independent crystallization of waxes by providing a lower kinetic route to crystallization compared to the normal wax nucleation in solution. It was also found that nucleating onto the PE plate involves an attenuated energy barrier compared to crystallizing along the normal nucleation route. The aggregate surface area was found to be of vital importance in the magnitude of the CFPP performance. This parameter increased as plate surface area increased. In turn, this latter parameter was found to scale with the segment molecular weights, i.e., as diblock molecular weights decreased plate surface area increased.

Another self-assembling butadiene-based system is the near-random copolymer containing ethylene and butene segments. The crystallinity can be tuned by the number of ethyl branches in the chain. For the series PEB-2 to PEB-12 the samples exhibit decreasing crystallinity while the PEB-13 to PEB-50 materials are wholly amorphous. The random PEB copolymers self-assemble to yield modulated rod-like structures. If the self-assembly event occurs prior to the wax phase separation temperature, the wax crystallization will be nucleated upon the self-assembled rod-like copolymer. Cocrystallization between the wax and the semi-crystallization segments of the copolymer can also occur when the self-assembly event is preceded by the wax crystallization event.

References

1. Radlinski AP, Barre L, Espinat D (1996) J Mol Cryst 51:383
2. Claudy P, Letoffe J-M, Bonardi B, Vassiladis D, Damin B (1993) Fuel 72:821
3. Coutinho JAP, Dauphin C, Daridon JL (2000) Fuel 79:607
4. Abdallah DJ, Sirchio SA, Weiss RG (2000) Langmuir 16:352
5. Venkatesan R, Nagarajan NR, Paso K, Yi YB, Sastry AM, Fogler HS (2005) Chem Eng Sci 60:3587
6. Denis J (1986) In: Bartz WJ (ed) 5th Intern Colloq Additive für Schmierstoffe, vol 2. Tech Ak, Esslingen
7. Richter D, Schneiders D, Monkenbusch M, Willner L, Fetters LJ, Huang JS, Lin M, Mortensen K (1997) Macromolecules 30:1053
8. Leube W, Monkenbusch M, Schneiders D, Richter D, Adamson D, Fetters LJ, Dounis P, Lovegrove R (2000) Energy Fuels 14:419
9. Monkenbusch M, Schneiders D, Richter D, Fetters LJ, Huang JS (1994) Il Nuovo Cimento 16:747
10. Schwahn D, Richter D, Wright PJ, Symon C, Fetters LJ, Lin M (2002) Macromolecules 35:861
11. Schwahn D, Richter D, Lin M, Fetters LJ (2002) Macromolecules 35:3762
12. Ashbaugh HS, Radulescu A, Prud'homme RK, Schwahn D, Richter D, Fetters LJ (2002) Macromolecules 35:7044
13. Radulescu A, Schwahn D, Richter D, Fetters LJ (2003) J Appl Cryst 36:995
14. Radulescu A, Schwahn D, Monkenbusch M, Fetters LJ, Richter D (2004) J Polym Sci B Polym Phys 42:3113
15. Radulescu A, Schwahn D, Stellbrink J, Kentzinger E, Heiderich M, Richter D (2006) Macromolecules 39:6142
16. Asbaugh HS, Fetters LJ, Adamson DH, Prud'homme RK (2002) J Rheol 46:763
17. Guo X, Pethica BA, Huang JS, Prud'homme RK, Adamson DA, Fetters LJ (2004) Energy Fuels 18:930
18. Ashbaugh HS, Guo X, Schwahn D, Prud'homme RK, Richter D, Fetters LJ (2005) Energy Fuels 19:138
19. Halperin A, Tirell M, Lodge TP (1992) Adv Polym Sci 31:100
20. Gast AP (1990) In: Scientific methods for the study of polymer, colloids and their applications. Kluwer Academic, Dordrecht, p 311
21. Tuzar Z, Kratochvil P (1976) Adv Colloid Interface Sci 6:201
22. Oranli L, Bahadur P, Riess G (1985) Can J Chem 63:2691
23. Bahadur P, Sastry NV, Marti S, Riess G (1985) Colloids Surf 16:337
24. Gallot Y, Franta P, Rempp P, Benoit HJ (1964) Polym Sci C473:4
25. Kotaka T, Tanaka T, Hattori M, Inagaki H (1978) Macromolecules 11:138
26. Periard J, Riess G (1973) Eur Polym J 9:687
27. Selb J, Gallot Y (1980) Makromol Chem 182:1491
28. Higgins JS, Dawkins JV, Maghami GG, Shakir SA (1986) Polymer 27:931

29. Plestil J, Baldrian J (1975) Makromol Chem 176:1009
30. Bluhm TL, Malhorta SL, Hong M, Noolandi J (1983) Polym Prepr (Am Chem Soc, Div Polym Chem) 24:405
31. Greenly RZ (1999) In: Brandup J, Immergut EH, Grulke EA (eds) Polymer handbook, 4th edn. Wiley, New York, p 309
32. Woffard CH, Hsieh HL (1969) J Pol Sci A1 7:461
33. Fetters LJ, Graessley WW, Krishnamoorti R, Lohse DJ (1997) Macromolecules 30:4973
34. Krisnamoorti R (1994) PhD Thesis, Princeton University, p 49
35. Hsieh HL, Randall JC (1982) Macromolecules 15:353
36. Choi SM, Barker JG, Glinka CJ, Cheng YT, Gammel PL (2000) J Appl Cryst 33:793
37. Koizumi S, Iwase H, Suzuki J, Oku T, Motokawa R, Sasao H, Tanaka H, Yamaguchi D, Shimizu HM, Hashimoto T (2006) Physica B 385–386:1000
38. Alefeld B, Dohmen L, Richter D, Brückel T (2000) Physica B 283:330
39. Kentzinger E, Dohmen L, Alefeld B, Rücker U, Stellbrink J, Ioffe A, Richter D, Brückel T (2004) Physica B 350:e779
40. Alefeld B, Schwahn D, Springer T (1989) Nucl Instrum Meth A 274:210
41. Alefeld B, Dohmen L, Richter D, Brückel T (2000) Physica B 276–278:52
42. Henke B, DuMond JWM (1953) Phys Rev 89:1300
43. Aschenbach B (1985) Rep Prog Phys 48:579
44. Alefeld B, Hayes C, Mezei F, Richter D, Springer T (1997) Physica B 234–236:1052
45. Debye P (1947) J Phys Colloid Chem 51:18
46. Schmidt PW (1991) J Appl Cryst 24:414
47. Beaucage G, Schaefer DW (1994) J Non-Cryst Solids 172:797
48. Porod G (1982) In: Glatter O, Kratky O (eds) Small-angle X-ray scattering. Academic, London, chap 2
49. Allen AJ (1991) J Appl Cryst 24:624
50. Ruland W (1987) Macromolecules 20:87
51. Müller G, Schwahn D, Springer T (1997) Phys Rev E 55:7267
52. Alexander S (1977) J Phys 38:983
53. deGennes PG (1980) Macromolecules 13:1069
54. Milner ST, Witten TA, Cates ME (1988) Macromolecules 21:2610
55. Pedersen JS (2000) J Appl Cryst 33:637
56. Dozier WD, Huang JS, Fetters LJ (1991) Macromolecules 24:2810
57. Hosemann R, Bagchi SN (1962) Direct analysis of diffraction by matter. North-Holland, Amsterdam
58. Machado ALC, Lucas EF (1999) Pet Sci Technol 17:1029
59. daSilva CX, Alvares DRS, Lucas EF (2004) Energy Fuels 18:599
60. Letoffe JM, Claudy P, Vassilakis D, Damin B (1995) Fuel 74:1830
61. Petinelli JC (1979) Rev Inst Fr Petr 34:791
62. Qian JW, Qi GR, Xu YL, Yang SL (1996) J Appl Pol Sci 60:1575
63. Qian JW, Zhou GH, Yang WY, Xu XL (2002) J Appl Pol Sci 83:815
64. Marie E, Chevalier Y, Eydoux F, Germanaud L, Flores P (2005) J Colloid Interface Sci 290:406
65. Zhang J, Wu C, Li W, Wang Y, Cao H (2004) Fuel 83:315
66. Wu C, Zhang J, Li W, Wu N (2005) Fuel 84:2039
67. Holder GA, Winkler J (1965) J Inst Petrol 51:228
68. Denis J, Durand J-P (1991) Rev Inst Fr Petr 51:637
69. Lewtas K, Tack RD, Beiny DHM, Mullin JW (1991) In: Garside J, Davey RJ, Jones AG (eds) Advances in industrial crystallization. Butterworth-Heinemann, London, p 166

70. Beiny DHM, Mullin JW, Lewtas K (1990) J Crystal Growth 102:801
71. Bennema P, Liu XY, Lewtas K, Tack RD, Rijpkema JJM, Roberts KJ (1992) J Cryst Growth 121:679
72. Clydesdale G, Roberts KJ, Lewtas K, Docherty R (1994) J Cryst Growth 141:443
73. Qian JW, Qi GR, Cheng RS (1997) Eur Polym J 33:1263
74. Qian JW, Wang X, Qi GR, Wu C (1997) Macromolecules 30:3283
75. Qian JW, Qi GR, Fang ZB, Cheng RS (1998) Eur Polym J 34:445
76. Chowdhury F, Haigh JA, Mandelkern L, Alamo RG (1998) Polym Bull 41:463
77. Goberdhan DG, Tack RD, Lewtas K, McAleer AM, Fetters LJ, Huang J (1996) Fuel oil additives and compositions, International Patent WO 96/28523
78. Schneiders D (1996) PhD Thesis, University of Aachen, Aachen, Germany
79. Monkenbusch M, Schneiders D, Richter D, Willner L, Leube W, Fetters LJ, Huang JS, Lin M (2000) Physica B 276–278:941
80. Mortensen K, Almdal K, Kleppinger R, Mischenko N, Reynaers H (1997) Physica B 241–243:1025
81. Takahashi Y, Noda M, Kitade S, Noda I (1999) J Phys Chem Solids 60:1343
82. Versmold H, Musa S, Dux Ch, Lindner P (1999) Langmuir 15:5065
83. Hamley IW (2000) Curr Opin Coll Interf Sci 5:341
84. Krisnamoorti R, Silva AS, Modi MA, Hammouda B (2000) Macromolecules 33:3803
85. Croce V, Cosgrove T, Dreiss CA, King S, Maitland G, Hughes T (2005) Langmuir 21:6762
86. Raphael E, deGennes PG (1992) Makromol Chem, Macromol Symp 62:1
87. deGennes PG (1979) Scaling concepts in polymer physics. Cornell University Press, Ithaca, NY
88. Vilgis T, Halperin A (1991) Macromolecules 24:3321
89. Elias HG (1990) In: Makromoleküle. Hüpf & Wepf, Basel, p 753
90. Gaucher V, Seguela R (1994) Polymer 35:2049
91. Gast AP, Leibler L (1986) Macromolecules 19:686
92. Buzza DMA, McLeish TCB (1997) J Phys II France 7:1379
93. Radulescu A, Schwahn D, Fetters LJ, Richter D (2002) Appl Phys A 74:s411
94. Radulescu A, Schwahn D, Monkenbusch M, Richter D, Fetters LJ (2004) Physica B 350:e927
95. Schwahn D, Yoo MH (1986) Springer Proc Phys 10:83
96. Brulet A, Lairez D, Lapp A, Cotton J-P (2007) J Appl Cryst 40:165
97. Schweins R, Huber K (2004) Macromol Symp 211:25
98. Benoit H (1953) J Polym Sci 11:507
99. Stellbrink J, Willner L, Jucknischke O, Richter D, Lindner P, Fetters LJ, Huang JS (1998) Macromolecules 31:4189
100. Beaucage G (1995) J Appl Cryst 28:717
101. Beaucage G (1996) J Appl Cryst 29:134
102. Barham PJ (1993) In: Thomas EL (ed) Materials science and technology, vol 12. VCH, Weinheim, p 153
103. Somani RH, Yang L, Zhu L, Hsiao BS (2005) Polymer 46:8587
104. Zwijnenburg A, Van Hutten PF, Pennings AJ, Chanzy HD (1978) Colloid Polym Sci 256:729

Editor: L. Leibler

Adv Polym Sci (2008) 210: 101–168
DOI 10.1007/12_2007_123

Published online: 25 October 2007

Layered Double Hydroxide Based Polymer Nanocomposites

Francis Reny Costa · Marina Saphiannikova · Udo Wagenknecht · Gert Heinrich (✉)

Leibniz-Institut für Polymerforschung Dresden e.V., Hohe Strasse 6, D-01069 Dresden, Germany
gheinrich@ipfdd.de

Abstract Nanocomposites based on polymers and inorganic filler materials not only create enormous interest among researchers because of their unique way of preparation and properties, but also promise development of new hybrid materials for specific applications in the field of polymer composites. The present article deals with the application of a relatively new class of inorganic materials, namely layered double hydroxides (LDHs), as nanofiller for synthesizing polymer-based nanocomposites. LDHs are mixed metal hydroxides of di- and trivalent metal ions crystallized in the form similar to mineral brucite or magnesium hydroxide (MH) with the incorporation of interlayer anionic species. Several procedures for the synthesis of LDHs, their organic modification, and the synthesis of polymer/LDH nanocomposites are discussed in detail with reference to work done in very recent years. The potential of LDHs, especially magnesium and aluminum-based LDHs (Mg – Al LDH) as nanofillers for the polymer matrix has been investigated. The important aspects in characterizing such hybrid materials (i.e., morphological analysis and melt rheological behavior) have been reported in detail to understand the nature of LDH particle dispersion and its influence on the melt flow behavior of the nanocomposites. The specialty of LDHs as nanofiller is their thermal decomposition behavior, which makes them potential flame retardants for polymers. This aspect has been reported in detail in the case of polyethylene-based systems, where the flame retarding efficiency of organically modified Mg – Al LDH alone and also in combination with conventional flame retardants has been discussed.

Keywords Layered double hydroxide · Nanocomposites · Rheology · Flame retardants · Memory effect

1 Introduction

In the last two decades, the synthesis and applications of nanocomposites based on polymers have drawn serious attention both from industry and academic sections. On the one hand, significant improvements in various properties (mechanical properties, gas permeability, electrical conductivity, photo and UV stability, controlled release, flammability, etc.) at low filler concentration are very attractive to industry. On the other hand, researchers show growing interest in developing new kinds of hybrid materials, studying the macromolecular dynamics under confined surroundings, and in investigating the influence of nanoscopic filler particles and their structural networks on materials properties such as dynamic mechanical, rheological, electrical, and thermal behavior. A number of review articles have already been published, highlighting the synthesis, characterization, and properties of polymer nanocomposites [1–4]. All these studies mostly concentrate on the use of layered silicate-type clay minerals as nanofillers in polymer matrices, where the organically modified clay particles are partly or fully dispersed in nanoscopic dimension (i.e., at least one dimension of the dispersed phase is in the nanometer range). At present, a different kind of layered inorganic material, generally called layered double hydroxide (LDH), is also gaining importance as nanofiller for the synthesis of polymer nanocomposites. These materials have a very wide range of chemical compositions based on different metal species, interlayer anions, etc. and are well known for their catalytic activities in organic synthesis [5]. The other potential applications of LDHs and their modified forms include biomedical applications e.g., controlled release of various drugs and biomolecules [6–8], gene therapy [8]), improvement of heat stability and flame retardancy of polymer composites [9–11], controlled release or adsorption of pesticides [12, 13], preparation of novel hybrid materials for specific applications (e.g., visible luminescence [14, 15], UV/photo stabilization [16, 17] and magnetic nanoparticle synthesis [18]), or in wastewater treatment [19, 20]. Compared to layered silicates, LDHs have the advantage of structural homogeneity, which can be tuned during their synthesis [21]. Additionally, high bound water content, non-toxicity, and high reactivity toward organic anionic species make them suitable for many specific applications. The idea of using LDHs as nanofiller for the synthesis of polymer nanocomposites is based on their two main characteristics: (i) their layered crystalline geometry with various intercalating anionic species, and (ii) their ability to interchange these interlayer anions with much larger organic anionic species. The latter is very important, as the pristine LDHs are not suitable for the penetration of giant polymer chains or chain segments into their gallery space unless the original interlayer distance is significantly increased through organic modification. Keeping this background in consideration, this article aims to present a brief overview of the various methods of synthesis of LDH materials and of the polymer nanocomposites based on LDHs. The synthesis and characterization of organically modified LDH are also discussed in detail for an understanding of their intercalation behavior in the presence of surfactant anions.

The LDH materials can be very interesting to industry as they combine the features of conventional metal hydroxide-type fillers, like magnesium hydroxide (MH), with the layered silicate type of nanofillers, like montmorillonite. The major area of interest in this regard is the role of LDH materials as potential non-halogenated, non-toxic flame retardant for polymer matrices. For years, scientists have been using the concept of nanotechnology to improve the flame retardancy of polymer nanocomposites. This approach involves the dispersion of inorganic filler, in nanoscale, as flame retardants into a polymer matrix. Usually, for this purpose, layered silicates and various other nanoparticles (MgO, MH, etc.) are used after suitable pretreatment. Several research reports have already shown that such an approach indeed improves the flame retardancy of the composites [22, 23].

Often a satisfactory improvement in flammability properties can be achieved in these composites in the presence of a small amount of secondary flame retardants. The advantages of this approach are the significant reduction in the amount of flame retardant loading and non-toxic decomposition products from the flame retardants. Like MH, Mg – Al LDH undergoes endothermic decomposition, thereby releasing the bound water and producing a metal oxide-type residue. This shows its potential as a flame retardant; it can inhibit or retard the flame propagation through three distinct means: (i) acting as heat sink through endothermic decomposition, (ii) forming a heat-insulating char on top of the burning surface, and (iii) diluting the volatile flammable gas through the release of bound water and carbon dioxide. Additionally, its layered silicate-like structure makes it suitable for achieving an improved dispersion of the filler particles through intercalation and exfoliation by polymer chains, which is unlikely in the case of MH. The major challenge in this regard is the homogeneous dispersion of an inorganic filler (like LDH) in a non-polar matrix (like polyolefins) through a melt compounding process. Therefore, a detailed study has been presented on polyethylene/Mg – Al LDH-based nanocomposites, highlighting the morphological and rheological behavior of the system to understand the nature of LDH particle dispersion and possible particle–particle and polymer–particle interaction. The role of LDH as flame retardant in such composites is also a matter of interest and hence is studied in detail.

2 Layered Double Hydroxide (LDH)

2.1 Structure

Layered double hydroxides (LDHs) belong to a general class called anionic clay minerals. They can be of both synthetic and natural origin. The most commonly known naturally occurring LDH clay is hydrotalcite, having the chemical formula $Mg_6Al_2(OH)_{16}CO_30 \cdot 4H_2O$. Hydrotalcite is the first mineral of this group whose structure and properties were studied extensively, and is often taken as the representative of the LDHs. Hence, LDHs are also known as hydrotalcite-like compounds. The general chemical formula of LDHs is written as $[M^{II}{}_{1-x}M^{III}(OH)_2]^{x+}(A^{n-})_{x/n}{\cdot}yH_2O$, where M^{II} is a divalent metal ion, such as Mg^{2+}, Ca^{2+}, Zn^{2+}, etc.; M^{III} is a trivalent metal ion, such as Al^{3+}, Cr^{3+}, Fe^{3+}, Co^{3+}, etc.; and A^{n-} is an anion, such as Cl^-, $CO_3{}^{2-}$, $NO_3{}^-$, etc. The anions occupy the interlayer region. Though a wide range of values of x is claimed to create a LDH structure, the pure phase of LDHs is usually obtained for a limited range, $0.2 \leq x \leq 0.33$ [24].

The structure of LDHs can best be explained by drawing analogy with the structural features of the metal hydroxide layers in mineral brucite or simply the MH crystal. A schematic representation comparing the brucite and the LDH structures is shown in Fig. 1. Brucite consists of a hexagonal close packing of hydroxyl ions with alternate octahedral sites occupied by Mg^{2+} ions. The metal hydroxide sheets in brucite crystal are charge neutral and stacked one upon another by Van der Waals interaction, resulting in a basal spacing of about 0.48 nm. In LDH, some of the divalent cations of these brucite-like sheets are isomorphously substituted by a trivalent cation. The mixed metal

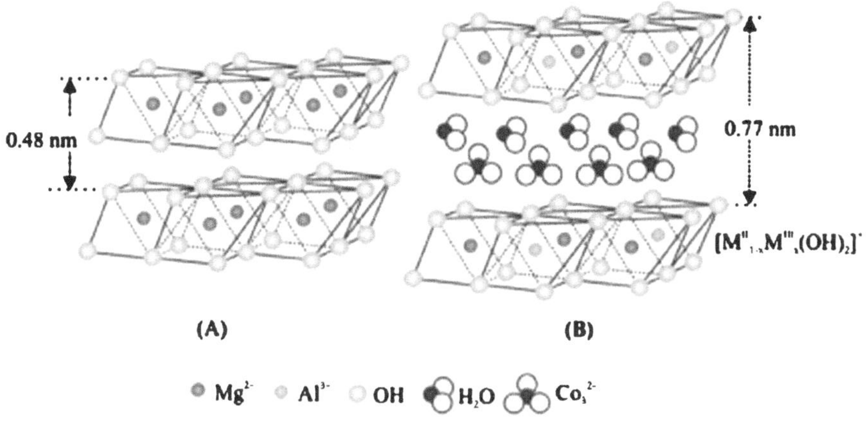

Fig. 1 Schematic representation comparing the crystal structure of brucite (*A*) and LDH (*B*)

hydroxide layer, i.e., $[M^{II}{}_{1-x}M^{III}{}_{x}(OH)_2]^{x+}$, thus formed acquires a net positive charge. This excess charge on the metal hydroxide layers is neutralized by the anions present in the interlayer region. The interlayer region in LDHs also contains some water molecules, as required for the stabilization of the crystal structure. The presence of anions and water molecules leads to an enlargement of the basal spacing from 0.48 nm in brucite to about 0.77 nm in Mg – Al LDH.

Although LDHs have a layered structure like the conventional layered silicate type of clays, these two materials are quite different from each other. While LDH has positively charged layers with anionic interlayer species (so they are called anionic clay), the layered silicates are of exactly opposite nature (hence called cationic clay). In terms of compositions, geometry, and layer thickness, the LDHs are vastly different from layered silicates. In LDH, as described earlier, each crystal layer is composed of a single octahedral metal hydroxide sheet whereas in layered silicates it is a sandwiched structure of two or more sheets of metal oxides. For example, a montmorillonite crystal layer is made up of three sheets: one octahedral sheet containing Fe, Al, Mg, etc. remaining sandwiched between two silica tetrahedral sheets. This difference in layer structure results in much lower crystal layer thickness and rigidity in LDH-type clays.

2.2 Synthesis

There are several methods by which LDHs can be synthesized. However, not all of these methods are suitable and equally efficient for every combination of metal ions.

The co-precipitation method involves the simultaneous precipitation of a selected pair of metal ions from their mixed aqueous solution by dilute NaOH and/or $NaHCO_3$, Na_2CO_3, or NH_4OH solution. The pH of the reaction medium is maintained in the range 8–10, depending on the nature of metal ions. Hydrothermal treatment of the final suspension is usually carried out to obtain well-crystallized samples. Detailed descriptions of the process are available in numerous reports available in the literature [25–29].

The homogeneous precipitation method using "urea hydrolysis" provides LDH with a high degree of crystallinity and a narrow distribution of particle size [30, 31]. Usually, an aqueous solution of desired metal ions and urea in proper molar ratio is heated in the temperature range from 90 °C to reflux temperature for 24–36 h. The urea molecules undergo decomposition producing ammonium carbonate, which finally causes the precipitation of LDH containing $CO^{2-}{}_3$ as the intercalating anion.

The hydrothermal crystallization method involves the crystallization of amorphous $M^{III}{}_2O_3$ precursor in the presence of a suitable $M^{II}O$, the latter acting as a crystallizing agent [32]. The precursor $M^{III}{}_2O_3$ is an amorphous hydrated oxide of the trivalent metal component of LDH whereas the crystallizing agent $M^{II}O$ is a reactive and basic oxide of the divalent metal component. The actual synthesis is carried out by hydrothermal treatment of an aqueous suspension of these two metal oxides in a pressurized vessel at elevated temperature for several days.

In the ion exchange method, the interlayer anions in LDHs are replaced by other anionic species. Using this method, LDHs containing different intercalating anionic species can be synthesized from one particular form [33–35]. However, the affinity of the anions to such interchange reaction depends on their charge and size. Usually, the original LDH is dispersed in an aqueous solution of the desired anionic species and the dispersion is stirred at room temperature for several hours.

2.3 Organic Modification

The modification of LDH materials is an inevitable step in the process of polymer nanocomposites preparation, especially when the melt compounding technique is employed. Since the hydroxide layers of all LDH clays are positively charged, the chemicals used for modification universally contain negatively charged functionalities. The primary objective of organic modification is to enlarge the interlayer distance of LDH materials so that an intercalation of large species, like polymer chains and chain segments, becomes feasible. Organic anionic surfactants containing at least one anionic end group and a long hydrophobic tail are the most suitable materials for this purpose. Due to these hydrophobic tails, the surface energy of the modified LDHs is reduced significantly compared to the unmodified LDHs. As a result, the thermodynamic compatibility of LDH with polymeric materials is improved, facilitating the dispersion of LDH particles during nanocomposite preparation.

Principally, the methods used for the synthesis of LDH materials are also suitable for their modification using organic surfactants. For example, methods like co-precipitation in aqueous solution of organic surfactants, ion exchange methods, etc. are widely reported in the literature [29, 36–38]. Besides all these methods, the "regeneration method" is also widely used for the modification of LDHs. This method is based on the "memory effect" shown by the carbonate-containing LDH materials. When such LDH is heated above 450 °C for several hours, it is converted into an amorphous mixed oxide, which after dispersion in an aqueous solution containing suitable anionic species and subsequent ageing, regenerates the original LDH structure (as if the materials remember its original structure) [39]. This regeneration is also possible in the presence of large anionic species, like many organic anionic surfactants. In the present work, Mg – Al LDH has been modified using the same principle. The intercalation behavior of Mg – Al LDH material was investigated with respect to different anionic surfactants, chosen according to their anionic functional groups and the nature of their hydrocarbon tail. The structure and chemical formula of these surfactant anions are given in Table 1. The length and nature of the hydrophobic tail of the surfactants are not identical. While dodecyl sulfate (DS) and laurate have the same tail of n-C_{12}, dodecylbenzene sulfonate (DBS) contains a benzene ring in the tail backbone, and bis(2-ethylhexyl)hydrogen phosphate (BEHP) contains two branched hydrocarbon chains. The unmodified LDH is first calcined in a muffle furnace for about 3 h at a temperature of 450 °C to convert it into a mixed oxide form (designated as CLDH). This mixed oxide is then dispersed in an aqueous solution of the desired surfactant and stirred for about

Table 1 Different surfactants used for intercalation within the LDH gallery [43]

Chemical name	Structure	Symbol of the anionic form
Dodecyl sulfate	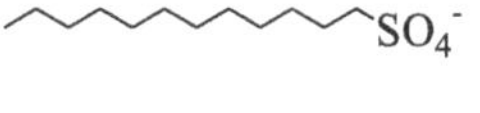	DS
Dodecyl benzenesulfonate	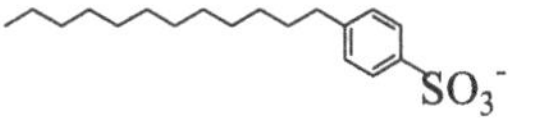	DBS
Laurate		Laurate
Bis(2-ethylhexyl) hydrogenphosphate	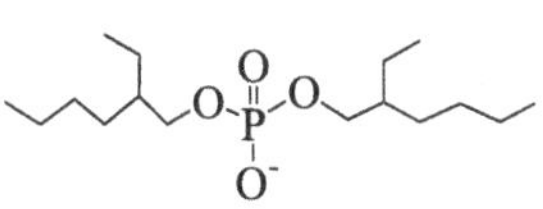	BEHP

24 h at ambient temperature. The concentration of the surfactant solution was maintained around 0.1–0.2 M and CLDH added to a specified volume of this surfactant solution in such an amount that there is enough surfactant anion available for 100% substitution of the interlayer carbonate anion after regeneration. The modified solid residue was then separated by centrifugation and dried at 60 °C to constant weight.

2.4 Characterization Modified LDH

2.4.1 X-Ray Diffraction Analysis

The X-ray diffraction (XRD) patterns of the surfactant-modified Mg – Al LDH are shown in Fig. 2. As expected, the position of the first order basal reflection ⟨003⟩ in all the modified samples is shifted to a higher *d*-value, indicating an expansion in the interlayer distance. Though none of the modified samples show distinct reflection at $d = 0.76$ nm, there appears a weak and broad reflection in the close vicinity. This can either be due to the presence of a small fraction of the unmodified LDH or to a higher order reflection in ⟨001⟩ series in the modified sample. However, the first option seems to be the most probable as, unlike the XRD pattern of the unmodified LDH, the modified samples contain reflections corresponding to a single ⟨001⟩ series. It also

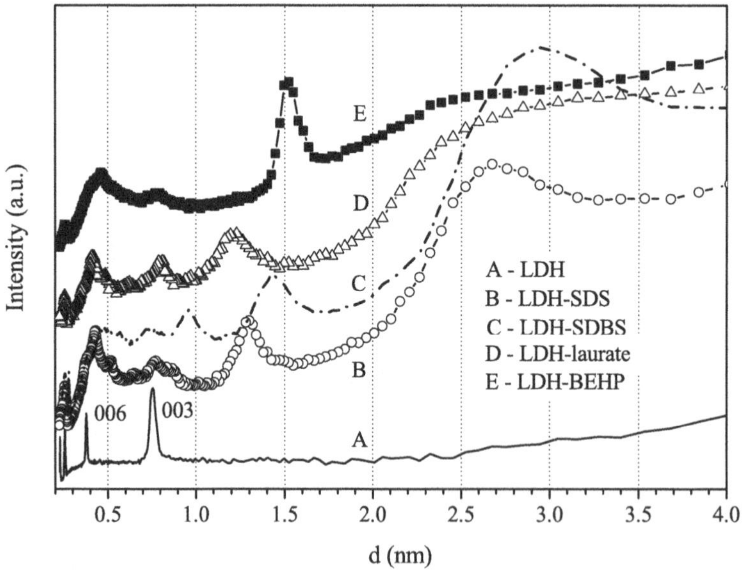

Fig. 2 XRD patterns of Mg – Al LDH and its various organically modified forms [43]

gives a strong hint of a loss in the crystalline order of the layered system, which may be due to the presence of only small crystallites and/or the loss of coherence conditions in certain crystallographic directions (i.e., no repeat units in the sense of scattering).

The presence of ⟨001⟩ reflections up to several orders in the modified samples indicates the regeneration of the crystal layers even in presence of large intercalating anions. The interlayer distance increases from 0.76 nm in the unmodified LDH to 1.52 nm in BEHP-modified LDH (LDH-BEHP), to 2.45 nm in laurate-modified LDH (LDH-laurate), to 2.68 nm in DS-modified LDH (LDH-DS) and to 2.95 nm in DBS-modified LDH (LDH-DBS). These values correspond well to those reported in the literature for these materials synthesized using the same surfactants and a similar or different chemical procedure. For example, LDH-DBS and LDH-DS synthesized by direct ion exchange reaction in aqueous solution of the surfactants results in an interlayer separation of 2.95 nm [40] and 2.62 nm [41], respectively.

2.4.2
FTIR Analysis

The Fourier transform infrared (FTIR) spectra of the modified LDH show two types of bands: one corresponding to the intercalated anionic species and the other corresponding to the host LDH material (Fig. 3). All the modified samples exhibit strong absorption bands in the range 2850–2965 cm^{-1}, corresponding to the $-CH_2-$ stretching vibration arising from the hydrocarbon tail present in each surfactant anion. The bands in the range 1000–1800 cm^{-1} are mostly due to the anionic functionalities present in the surfactant and also due to interlayer water molecules. The band around 428 cm^{-1} is due to the lattice vibration of the hydroxide sheet, and the broad band in the range 3200–3700 cm^{-1} is mainly due to $O-H$ groups present in metal hydroxide layers. The appearance of characteristic bands for CO_3^{2-} (γ_2, γ_3, and γ_4) means that some CO_3^{2-} still exists in the interlayer region of the modified materials. However, the absence of the strong 1357 cm^{-1} band in these samples is an indication of significant decrease in the interlayer CO_3^{2-} content and change of its symmetry. The lowering of CO_3^{2-} ion symmetry in the interlayer region is further indicated by the splitting of the γ_3 peak into a pair at around 1379 cm^{-1} and 1467 cm^{-1} [30, 42]. This is perhaps caused by partial release of the constraint on movement of the CO_3^{2-} ions due to enlargement of interlayer region after organic modification.

In addition to the carbonate ion not released during the calcination process, there can be incorporation of fresh carbonate ions during the regeneration process from the atmospheric CO_2 dissolved in the dispersion medium. The FTIR spectra of the modified samples do not provide any clear indication of the presence of interlayer water. In this regard, the only difference observed is the disappearance of a weak band (in the form of a shoulder) in

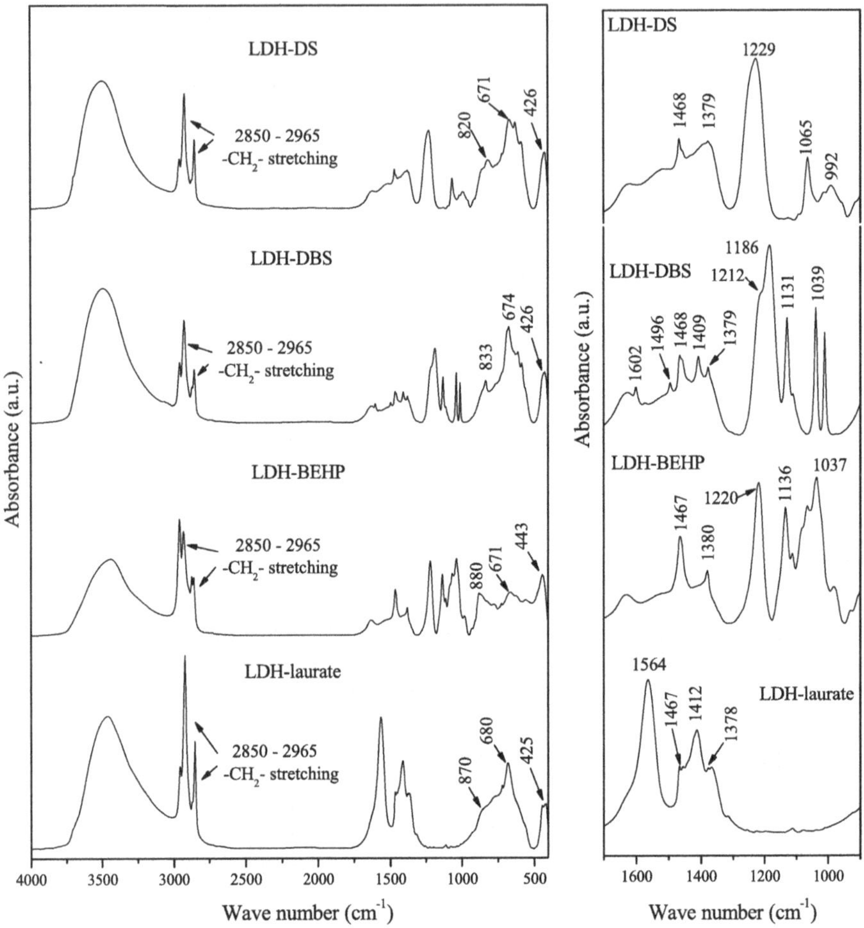

Fig. 3 FTIR spectra of LDH modified with different anionic surfactants by the regeneration method [43]

the region 3000–3100 cm^{-1}, which originates from the interaction between H_2O molecules and CO_3^{2-} ions. However, a broad band or shoulder is observed in the modified samples in the region 1600–1640 cm^{-1}, which may be due to the bending vibration of H_2O. However, the XRD and thermogravimetric analysis (TGA) analyses indicate the presence of water molecules in the interlayer region [43]. Therefore, it seems logical to interpret that the water molecules present in the modified LDH do not interact with CO_3^{2-} anions, rather they bridge the gap between the hydrocarbon tail of the surfactants and the metal hydroxide layers.

Further, anionic functional groups of the surfactants also exhibit their characteristic FTIR bands. In LDH-DS, the characteristic $S=O$ stretching vibration bands appear at 1229 cm^{-1} (symmetric) and 1065 cm^{-1} (asymmetric). Whereas the corresponding bands in LDH-DBS appear at 1186 cm^{-1} and 1038 cm^{-1}, respectively. The $C-S$ stretching vibration band is also observed at 630 cm^{-1} in LDH-DS and at 615 cm^{-1} in LDH-DBS. The LDH-DBS additionally shows multiple bands corresponding to the aromatic ring $C=C$ vibrations in the range 1450–1610 cm^{-1}. The LDH-BEHP exhibits characteristic $P-O-C$ stretching vibration bands at 1136 cm^{-1} (symmetric) and 1037 cm^{-1} (asymmetric). The $P=O$ stretching vibrations is indicated by a strong band at 1220 cm^{-1}. The spectrum of LDH-laurate has two strong characteristic bands at 1563 cm^{-1} and 1412 cm^{-1}, respectively, for the asymmetric and the symmetric stretching vibrations associated with the COO^{-1} group [44]. The intercalation imparts some degree of constraint on the various characteristic vibrations of these functional groups. As a result, their corresponding FTIR bands are expected to shift to lower values of wave number in comparison to their free-state values, as more energy is required for executing such vibrations in the presence of constraints.

2.4.3 Morphological Analysis

Synthetic $Mg-Al$ LDH has usually plate-like particle morphology. The size distribution of the particles depends mostly on the synthesis conditions and varies from few hundred nanometers to few micrometers in lateral dimensions. In Fig. 4a, the SEM micrograph of the synthesized LDH reveals mostly a plate-like particle geometry with a distinct hexagonal shape and sharp edges. The highly anisometric nature of these primary particles is also apparent. The lateral dimension of these particles varies within few micrometers, whereas the thickness hardly exceeds few hundred nanometers. Interestingly, the calcination at about 450 °C does not change the general particle morphology of the synthesized $Mg-Al$ LDH. As Fig. 4b shows, the plate-like appearance of the primary particles still persist in calcined LDH. The morphological features of the modified LDH are quite similar, irrespective of the type of surfactant used in the present study. The regeneration process restores the metal hydroxide sheets of the LDH crystal. However, the particle morphology is slightly changed after organic modification. As can be observed in Fig. 4c–f, the well-defined hexagonal particle shape is lost. Instead, plate-like particle morphology with irregular shapes and edges predominates in the modified LDH. This seems quite obvious as the regeneration process, even in presence of carbonate anion, does involve a loss of crystallinity. The surfactant anions being much larger in size than simple inorganic anions perhaps hinder the large-scale lateral growth of the LDH layer. All the modified samples except LDH-BEHP show significant surface irregularities compared to the unmodified sample.

Careful observation of the higher magnification scanning electron microscope (SEM) images reveals that the surface texture of primary par-

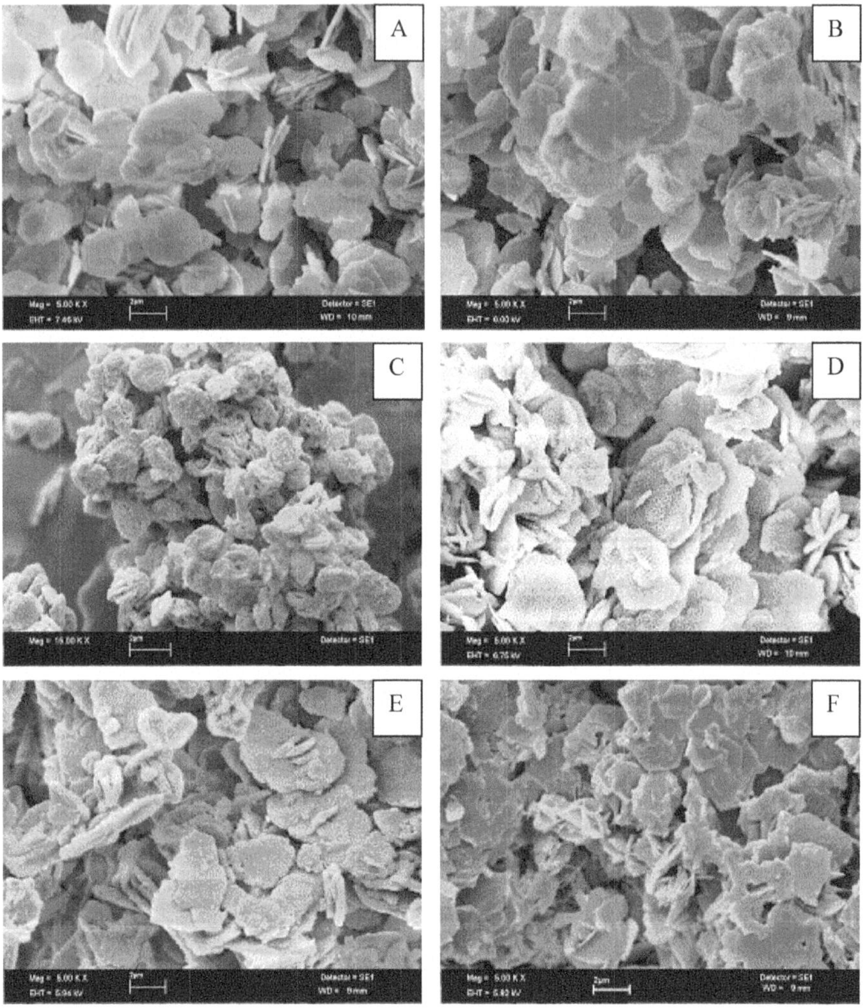

Fig. 4 SEM micrographs of the LDH samples: **a** unmodified LDH, **b** calcined LDH, **c** LDH-laurate, **d** LDH-DS, **e** LDH-DBS and **f** LDH-BEHP (*magnification bar* is 2 μm) [43]

ticles (platelets) of the three samples (LDH-laurate, LDH-DBS and LDH-DS) is different from that of LDH-BEHP. In can be seen from Fig. 5 that in these three modified LDH samples, the particle surface is either perforated or contains some secondary layer growth. Also, the particles appear more floppy compared to unmodified LDH. In contrast, LDH-BEHP has quite similar surface texture to the unmodified LDH, which has a smooth surface and sharp edges. The structure of BEHP is quite different from that of the other surfactants. Though it contains two hydrocarbon tails, the length of each tail is much smaller than those present in the other three surfac-

Fig. 5 Appearance of the surface morphology of the primary particles in LDH-laurate (**A**), LDH-DS (**B**), and LDH-DBS (**C**) (*magnification bar* is 2 μm) [43]

tants. The XRD analysis also reveals that in LDH-DS and LDH-DBS the expansion of the interlayer distance is much higher than that in LDH-BEHP. The size of the surfactant anions may be a potential factor that influences stacking and growth of the metal hydroxide layers during the regeneration process. However, further investigations are necessary for determining the exact mechanism of the regeneration process in the presence of organic surfactants.

3 Preparation of LDH-Based Polymer Nanocomposites

3.1 In-Situ Methods

3.1.1 In-Situ LDH Synthesis

In this method, LDH is synthesized in situ in a polymer solution and during the stacking of LDH layers polymer chains intercalate between the layers. Typically, the two constituent metal salts are dissolved in a desired polymer solution first, followed by the co-precipitation of the metals in the double hydroxide form. Messersmith and Stupp [45] prepared poly(vinyl alcohol) (PVA) intercalated Ca – Al LDH by precipitating the latter in the presence of dissolved PVA. The process involved mixing of a $Ca(OH)_2$/PVA solution to a solution containing $Ca(OH)_2$ and $Al(OH)_3$. It was suggested by them that the PVA molecules facilitate the nucleation and the growth of the metal hydroxide layers. Oriakhi and coworkers [46] intercalated poly(acrylic acid), poly(vinyl sulfonate), and poly(styrene sulfonate) into LDHs such as $M_{1-x}Al_x(OH)_2{}^{x+}$ (M is Mg, Ca, Co) and $Zn_{1-x}M_x(OH)_2{}^{x+}$ (M is Al, Cr) by precipitating LDH from a deareated aqueous basic solution containing mixed metal nitrates and dissolved polymer. Other examples of polymer/LDH nanocomposites prepared by direct intercalation through the co-precipitation technique are Mg – Al LDH intercalated with poly(styrene sulfonate), poly(vinyl sulfonate) [47], polyaspartate [48], and poly(ethylene oxide) derivatives [49], etc.

3.1.2
In-Situ Polymerization

In-situ polymerization is the most widely referred technique for preparation of LDH-based polymer nanocomposites. This process is solution-based and is usually carried out in an aqueous system. The scheme shown in Fig. 6 indicates the general principle for carrying out in-situ polymerization within the layers of LDH crystals. The primary step in this procedure is the preparation of monomer intercalated LDH hybrids, which are then subjected to excitation using heat [48, 50], initiating chemicals [51], etc. to initiate the polymerization reaction. Various methods of intercalation of monomers into the interlayer space of LDH crystals and their subsequent polymerization have been reported in literature and are summarized in Fig. 6.

The anionic interchange method involves the dispersion of LDH materials into a monomer solution, most often in aqueous medium (path 1 in Fig. 6) [52–54]. The dispersion is then stirred for several hours with mild heating. To be able to replace the interlayer anion, the monomer molecules should have such functionality as can stabilize the layered structure by neutralizing the excess charge on the hydroxide sheets of LDH. For example, acrylate anions are intercalated into Mg – Al LDH through ion exchange with Cl^- or $NO_3{}^-$ present in LDH. Lee and Chen reported intercalation of acrylate and 2-acryloamido-2-methyl propane sulfonate into hydrotalcite [51]. These intercalated hybrids are then dispersed in an alkali-neutralized solution of the monomer and the polymerization is carried out in the presence of an initiator. Leroux and coworkers prepared vinylbenzene sulfonate monomer intercalated Zn – Al LDH [52] and aminobenzene sulfonate monomer intercalated Cu – Cr LDH [53]. Further, Tanaka et al. reported acrylate ion intercalation within LDH by replacing the nitrate anion from Mg – Al LDH [54].

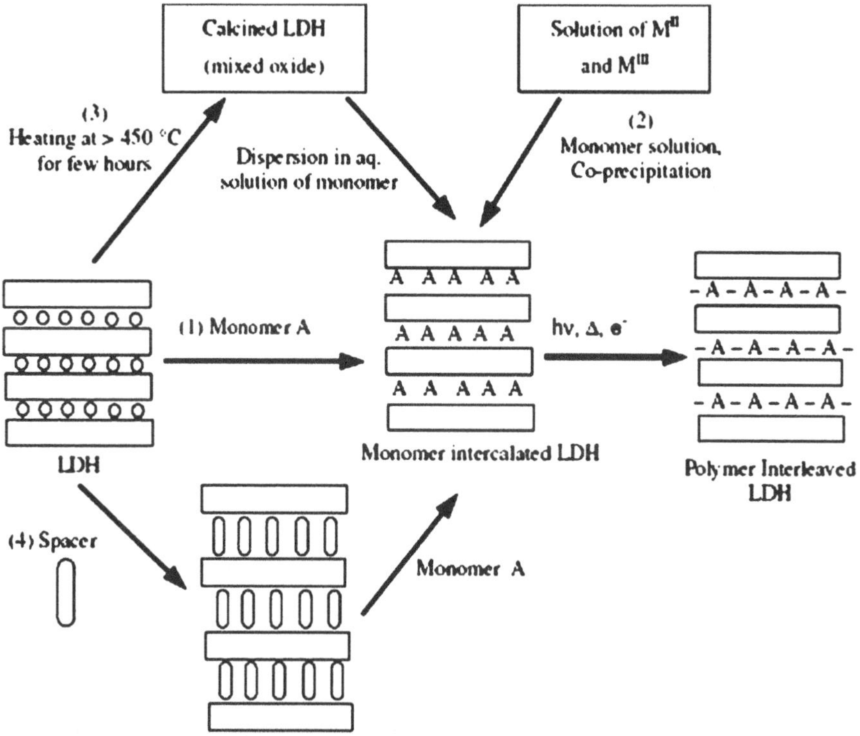

Fig. 6 Schematic pathways of in-situ polymerization within LDH layers to synthesize polymer/LDH nanocomposites

Further, Isupov et al. investigated the intercalation of various isomeric anions of aminobenzoic acid into Li – Al LDH and their subsequent polymerization [55].

The most common and successful method of preparation of monomer–LDH hybrid is the synthesis of LDH by co-precipitation of metal ions from their mixed solution containing dissolved monomer (usually in the form of salt) (path 2, Fig. 6). To minimize interference with CO_3^{2-}, the reaction mixture is often purged with nitrogen. The formation of hydroxide layers and the inclusion of monomer anion in the interlayer region take place simultaneously. Whilton and coworkers prepared poly(amino acid)/LDH-based nanocomposites following this method [48]. An amino acid intercalated Mg – Al LDH was prepared by reacting a mixed Mg and Al nitrate solution with a basic solution containing aspartate anion under nitrogen atmosphere. The aspartate–LDH hybrid thus obtained was subjected first to heating at 220 °C for 24 h followed by treatment with basic solution. The heat treatment in the first step provides condensation of aspartate monomer within the interlayer region into polysuccinimide, which in the second step undergoes hydrolysis to form poly(α,β-aspartate) [48]. Other examples of synthesis of polymer/LDH nanocomposites by co-precipitation methods are intercalation and subsequent polymerization of styrene-4-sulfonate within Ca – Al LDH [56], styrene sulfonate into Zn – Al and Ca – Al LDH [57], and 3-sulfopropyl methacrylate within Zn – Al LDH [58], etc.

The ability of LDH materials to regenerate their layered structure from an aqueous dispersion of their mixed-oxide form is also applied to prepare the monomer-intercalated hybrids (path 3, Fig. 6). This is called the regeneration method, which is similar to the procedure used for converting CO_3^{2-}-containing LDH to other forms.

The organic/inorganic pillaring method (path 4, Fig. 6) differs from the anion exchange method in the way that an intermediate anionic species or pillaring agent is used in this method as spacer to increase the interlayer distance before the intercalation of monomer. Therefore, the monomers in this case interact with a pillared LDH species. Wang et al. [59] used 10-undecenoate pillared Mg – Al LDH hybrid (prepared by the co-precipitation method) as the precursor for the intercalation of methyl methacrylate (MMA) into the LDH layer and subsequent polymerized MMA using a two-step bulk polymerization technique. In the first step, called the prepolymerization step, a homogeneous mixture of organically pillared LDH and MMA monomer was prepared through constant stirring, which resulted in intercalation of MMA into the LDH layer. The mixture was then heated at 50 °C under nitrogen atmosphere in the presence of a small amount of catalyst 2,2′-azobisisobutyronitrile (AIBN) to prepolymerize the MMA monomers. In the second step, called the casting polymerization, AIBN was again added to the prepolymer mixture at room temperature and injected into a glass mold at 60 °C for 4 h. The glass mold was then kept in an oven at 120 °C for 1 h to complete the polymerization process. In another variation, amino benzoate pillared LDH was used to prepare poly(methyl methacrylate) (PMMA)/LDH nanocomposite using a single-step solution polymerization technique [60]. Challier and Slade [61] carried out intercalation of aniline molecules into terephthalate or hexacyanoferrate pillared Cu – Cr and Cu – Al LDH. The interlamellar oxidative polymerization of aniline molecules yields polyaniline intercalated LDH with retention of the framework of the latter. Sugahara and coworkers [62] intercalated acrylonitrile monomer into a dodecyl sulfate pillared Mg – Al LDH to synthesize polyacrylonitrile in the interlayer region. Similarly, O'Leary and coworkers [63] also used dodecyl sulfate pillared Mg – Al LDH to intercalate acrylate monomer and its subsequent polymerization. Li et al. [64] reported the intercalation polymerizable acrylamide monomer into a glycine-pillared Mg – Al LDH. The subsequent heating of such monomer intercalated LDH hybrid at about 120 °C resulted in polymerization of the acrylamide monomers in the interlayer region. Hsueh and Chen [65] introduced a slight modification in this method to synthesize LDH/polyimide nanocomposites. Amino benzoate-modified Mg – Al LDH was used as the precursor for intercalation of monomers (pyromellitic anhydride and 4,4′-oxydianiline). The formation of polyimide in the interlayer region takes place in two steps through the formation of polyamic acid. The possible mechanism of interlayer polymerization is shown in Fig. 7. The difference here is that the modifying anions (aminobenzoate) are attached to the polymer chain ends and cause their anchorage to LDH sheets through ionic interaction.

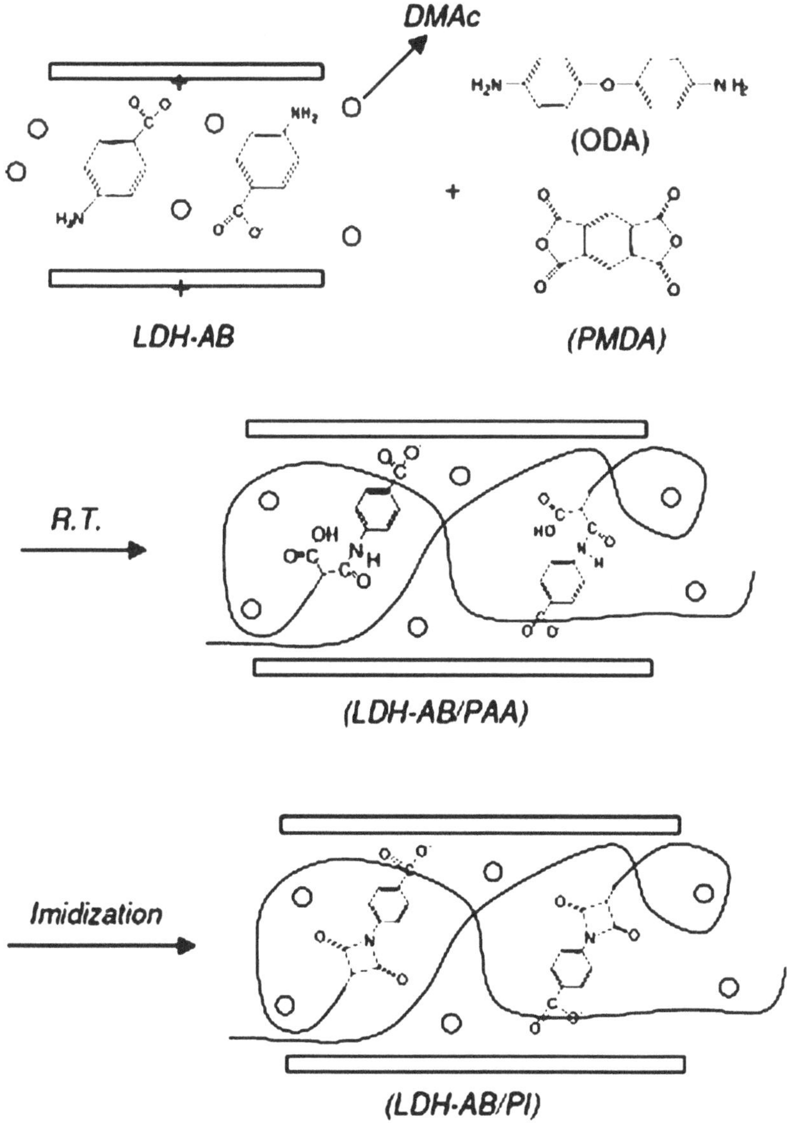

Fig. 7 Preparation of LDH-polyimide (LDH-AB/PI) nanocomposites. *LDH-AB* aminobenzoate intercalated LDH, *ODA* 4,4′-oxydianiline, *PMDA* pyromellitic anhydride (Reprinted from [134] with permission from Elsevier)

3.2 Solution Intercalation

The direct intercalation of the polymeric species having functionalities that can interact with the hydroxide layers of LDH is also an effective way to prepare polymer/LDH nanocomposites. Principally, the methods used for intercalation of monomer or small oligomeric organic species in the interlayer region of LDH, are also applicable for high molecular weight species. But, due to very small interlayer distance and the high charge density of LDH layers, the direct intercalation of large polymeric chains is much more difficult. Therefore, instead of pristine LDH, organically modified LDHs are often used in this method. Usually, organically modified LDH is first dispersed in a solution of the desired polymer, followed by ageing with stirring under nitrogen atmosphere to accomplish polymer intercalation.

Qu and coworkers prepared and characterized polyethylene/LDH [66, 67] and polystyrene/LDH nanocomposites [68] by this method. They used DS-modified LDH prepared by the reconstruction and anion exchange method. The nanocomposites were obtained by refluxing the mixture of DS-modified LDH and a xylene solution of the polymers. Buniak and coworkers [41] also used DS-modified Mg – Al LDH to prepare poly(ethylene oxide)/LDH nanocomposites. The modification of LDH was carried out using ion-exchange method and the modified LDH was later treated with aqueous solution of poly(ethylene oxide) to prepare the nanocomposite. Recently, Liao and Ye [69] prepared poly(ethylene oxide)/LDH-based nanocomposite electrolyte by the solution intercalation method. They modified Mg – Al LDH using the co-precipitation method with two different types of anionic surfactants: one was *n*-alkyl 3-sulfopropyl ether salt (containing about 11 ethylene oxide/mole and C_{13}–C_{15} alkyl chain) and the other oligomeric hydroxy polyethylene oxide phosphate. The nanocomposites were obtained by mixing poly(ethylene oxide), modified LDH, and lithium perchlorate in anhydrous acetonitrile at 70 °C with constant stirring for several hours. Hsuesh and Chen [70] prepared epoxy/LDH nanocomposites by treating amino laurate-intercalated Mg – Al LDH with a mixture of epoxy resin and curing agent. The resin and the curing agent diffused into the LDH interlayer region after heating the mixture for few hours. After thermal aging exfoliated epoxy/LDH nanocomposites were obtained. Further, Li and coworkers [71] used glycine-intercalated Mg – Al LDH and modified it with formamide, which was then treated with an acetone solution of PMMA to form the nanocomposites.

The use of unmodified LDH for direct anion exchange with polymeric materials has also been reported. For example, Yang and coworker [72] carried out intercalation of poly(ethylene oxide) sulfate and poly(ethylene glycol) within Mg – Al LDH by aging a polymer/LDH aqueous suspension at 65 °C for 4 days. Costa and coworkers [73] prepared nanocomposites through intercalation of dendrimers (carboxylate-terminated polyamidoamide) into LDH.

They observed saturation of dendrimer intercalation into the host LDH when the mixing ratio of the two ingredients exceeded 1 : 2 (charge ratio of anionic clay and carboxylate group of the dendrimer). The dendrimers remained densely packed in the interlayer region of LDH and had an ellipsoidal shape. When excess dendrimers were used in solution for intercalation (mixing ratio 1 : 8), in addition to intercalation they were also adsorbed on the clay surface. Similarly, Leroux and coworkers [74] synthesized poly(ethylene oxide) derivative-intercalated Cu – Cr LDH nanocomposites by replacing the Cl^- anion from the interlayer region. They first ultrasonicated the LDH material in decarbonized water, which was then added to a neutralized polymer solution containing an amount of polymer equivalent to twice the anion exchange capacity of the clay.

3.3 Melt Compounding Method

Perhaps the most challenging method of preparing polymer/clay nanocomposites is the melt compounding method. With non-polar polymers, like polyolefins, this method becomes more difficult due to high thermodynamic incompatibility between the non-polar matrix and the polar clay materials. Application of the melt compounding method has definite technological advantages over the solution method as the former can be easily adopted for industrial product manufacture using conventional polymer processing equipment. Therefore, the melt compounding method always finds special interest among researchers. Use of Mg – Al LDH or hydrotalcite as filler in the preparation of polymer composites using this method is very common, where it is treated as simple metal hydroxide whose basic nature and endothermic decomposition serve a twofold purpose: as acid scavengers in halogenated polymers [9,75], and as flame retardant [10]. However, in any of these applications no reference was made to exploit the efficiency of Mg – Al LDH as nanofiller.

The melt compounding method used for the preparation of polymer/LDH nanocomposites is similar to that used for conventional polymer/clay nanocomposites. The LDH clay materials are first modified by an organic surfactant using the methods described earlier. The organically modified LDH is then mixed with molten polymer in the typical plastic processing equipments. Nichols and coworkers [76] were the first to report the melt compounding method for preparing polymer/LDH nanocomposites. Recently, researchers have started to show more interest in this method to prepare polymer/LDH-based nanocomposites using different types of matrices. Costa et al. [77] prepared LDPE/Mg – Al LDH-based nanocomposites, where a master batch of dodecyl benzenesulfonate-modified LDH in maleic anhydride grafted polyethylene was diluted in a base LDPE resin using a two-step melt compounding process. For polyolefin-based nanocomposite systems this two-step method is very common. In the case of a polar polymer, the single step melt compounding process is used, where modified LDH is directly compounded with base polymer resin. For example, Zammarano et al. [78] and Du et al. [11] prepared polyamide 6/LDH nanocomposites using organically modified Mg – Al LDH. It has been observed that a high degree of exfoliation of the LDH particles can be obtained using LDHs with low anion exchange capacity and at reasonably high LDH loading (for example up to 10 wt %. Lee et al. [79] prepared poly(ethylene terephthalate)/LDH nanocomposites with Mg – Al LDH modified with various organic surfactants. They observed improved thermal and mechanical properties of these nanocomposite compositions compared to those of the unfilled polymer. The other reported polymer/LDH nanocomposite systems prepared by melt compounding methods include polypropylene/Zn – Al LDH [80], polyethylene/Mg – Al LDH [81], polyethylene/Zn – Al LDH [82], etc.

4 Polyethylene/Mg – Al LDH Nanocomposites

4.1 Background

Among all the metal ion combinations, the Mg – Al pair is one of the most widely reported for polymer nanocomposite preparation. The specialty of the LDHs based on this ion pair is their natural occurrence (only variety of LDH), biocompatible nature, comparatively more easy synthesis in highly crystalline and small particle-sized form, etc. Also, the thermal decomposition behavior of these materials is between those of MH and $Al(OH)_3$, the two widely used flame retardant inorganic fillers for polyolefins, polyesters, etc. In fact, one of the primary objectives for the use of Mg – Al LDH as nanofillers for polymer is to explore its potential as flame retardant for the polymer matrix. The major driving force in this regard is to develop a non-toxic and environmentally friendly flame retardant. In the case of polyolefins, these materials find special interest in improving the dispersion of metal hydroxide-type filler through the intercalation–exfoliation process involved during polymer/clay nanocomposite compounding. Such a mechanism is not practicable with conventional metal hydroxides like MH and $Al(OH)_3$, but provides an effective way to improve the dispersion and hence the efficiency as flame retardants.

Low-density polyethylene/Mg – Al LDH (LDPE/LDH) nanocomposites were prepared in a tightly intermeshing co-rotating twin-screw extruder having 27 mm screw diameter and L/D ratio value of 36. In the first step, DBS-modified Mg – Al LDH (LDH-DBS) and maleic anhydride grafted polyethylene (PE-g-MAH) were compounded in the extruder in 1 : 1 weight ratio. The extruded masterbatch was then water-cooled as it emerged from the extruder die and granulated. The masterbatch was then dried at 60 °C for about 2 h. In the second step, the dried masterbatch granules and desired amount of LDPE were premixed and compounded in the same extruder. The processing conditions used for both these steps are as follows:

- 160–210 °C temperature profile from the feed to the die section of the extruder barrel
- 200 rpm screw speed
- 6 kg h^{-1} feed rate
- Vacuum outlet in the mixing section of the extruder barrel to take out any volatiles formed during compounding

The final composites as extruded were water-cooled, granulated and dried at 60 °C for 2 h. It has been estimated that the LDH-DBS contains about 46 wt % metal hydroxide. The concentration of filler in the final composites was determined considering the metal hydroxide content of the modified LDH. They are designated as PE-LDH1 (2.43 wt % modified LDH), PE-LDH2 (4.73 wt %), PE-LDH3 (6.89 wt %), PE-LDH4 (8.05 wt %), PE-LDH5 (12.75 wt %), and PE-LDH6 (16.20 wt %).

All the ingredients used for the nanocomposite preparation are commercially available and were used without further purification.

4.2 Morphological Characterization

A rigorous morphological characterization of a polymer/clay nanocomposite is necessary to understand the overall state of clay particle dispersion both at micro- and nanoscales in the polymer matrix. The primary approach is to study the changes in the interlayer separation of the clay crystals using XRD at various stages of material preparation. In the case of melt compounding processes, such analysis is usually carried out with freshly compounded nanocomposite samples and the results are compared with those observed in the clay precursor, i.e., organically modified clay. However, the XRD analysis alone is not reliable for drawing a final conclusion about the dispersion state of clay particles in nanocomposites [83, 84]. This is because of the fact that the nature of reflection maxima (position and intensity) in an XRD pattern not only depends on the interlayer separation, but also on several other factors, like concentration of the clay in the nanocomposites, symmetry in a specific crystallographic direction, etc. Besides, inhomogeneity in the crystal structure of the clay (i.e., intercalated, exfoliated, and unmodified fraction existing in the same sample) may result in a complicated XRD pattern. Therefore, along with XRD analysis, electron microscopic investigations are used as complementary techniques to obtain a complete picture of the clay particle dispersion state in polymer nanocomposites. However, the morphological analysis using electron microscopy techniques (TEM, SEM, etc.) have certain inherent disadvantage when polymer nanocomposite characterization is concerned. First, to view nanoscopic particle distribution within a polymer matrix one needs to rely on high magnification images, which makes the scan surface area very small. Thus, it does not necessarily represent the whole surface area of the sample. Second, it shows only the particle distribution in the surface layer, unless the sample test samples are prepared from the bulk region. In the case of molded articles or test specimens, the nature of particle distribution can be significantly different at the different sections of the samples [85]. To overcome the small scan area problem, images of different magnifications, starting from low magnification to high magnification, should be considered to obtain a clear picture of the nature of particle distribution both at macro- and nanoscale.

4.2.1 X-Ray Diffraction Analysis

The wide-angle X-ray spectroscopic (WAXS) patterns of several LDPE/Mg – Al LDH nanocomposites are shown in Fig. 8. The WAXS patterns of all the compositions contain the first order basal reflection (⟨003⟩), indicating that the LDH layers are not fully exfoliated in the LDPE matrix. Another interesting point to be noted here is the nature of the higher order basal reflection of ⟨001⟩ series. In comparison to LDH-DBS, they are either disappeared or their intensities are sharply reduced. This is a clear indication of increasing crystal disordering of the LDH-DBS particles in the nanocomposites [84].

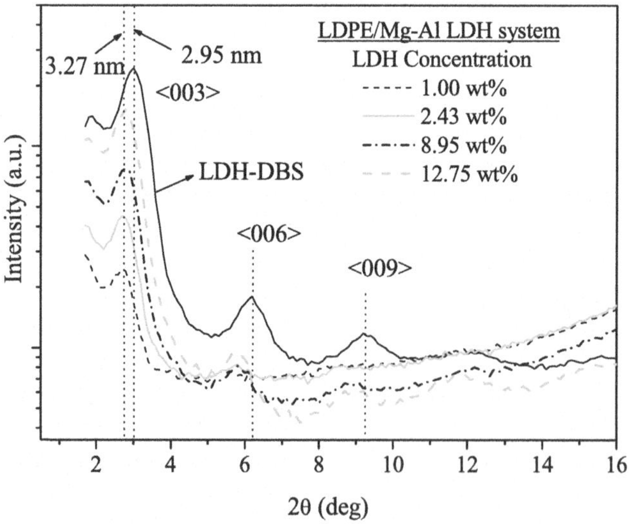

Fig. 8 WAXS patterns of LDPE/Mg – Al LDH compositions

The first order basal reflection in LDH-DBS corresponds to an interlayer layer separation of about 2.95 nm. In the nanocomposite, this reflection shifts to a slightly higher value of 3.27 nm. This may be due to the intercalation of polymer chain-segments into the interlayer region of LDH, which is induced by a strong shearing during melt mixing in the extruder. From the WAXS pattern it is apparent that the LDPE/Mg – Al LDH prepared by melt compounding does not show fully exfoliated structure or a high degree of exfoliation of the dispersed LDH particles.

4.2.2
TEM Analysis

More direct evidence of filler particle morphology in the LDPE/Mg – Al LDH nanocomposite can be obtained from a systematic analysis of the TEM images of these materials. The changes in the morphological features of the LDH-DBS particles induced by the melt compounding process in extruder are shown in Fig. 9. The SEM image shown in Fig. 9a reveals the nature of LDH-DBS particles, which roughly consists of plate-like shape with lateral dimensions ranging over few a micrometers and thickness over few hundreds of nanometers. These primary particles during melt compounding suffer significant size reduction, as is obvious from the TEM images of various nanocomposites (Fig. 9b–e). These low magnification images cannot reveal nanoscopic particle structures in the matrix, but provide a direct way of inspecting the presence of primary LDH-DBS clay particles or their aggregated structures. At first look, the discrete filler particles can be detected throughout the matrix, whose dimensions falling roughly within the 1 μm scale. These particles can be either individual primary particles (platelets/tactoids consisting of multiple metal hydroxide layers) when viewed perpendicular to their lateral face or clusters of primary particles when viewed parallel to the lateral surface. With increasing filler concentration, the size of these particles remains practically unchanged, only their number density increases. Even at an LDH concentration of 12.75 wt %, these discrete particles does not form large aggregates to an appreciable extent. It is also apparent that a strong shearing during melt compounding causes significant rupture of the primary LDH-DBS particles.

When the high magnification TEM images are analyzed, both the nature of the large LDH particles (in Fig. 9b–e) and the matrix regions in between them become more distinguishable. Figure 10 gives a magnified view of one such particle and also the surrounding matrix region. It is apparent that these large particles are highly distorted and show ample evidences of crystal layer delamination at the surface. Such particles will always exhibit poor long range crystalline symmetry and this explains the broadening and weakening of reflection maxima in the XRD patterns described in Fig. 8. The morphological features observed in Fig. 10b give a clear understanding of the mechanism of exfoliation of the LDH-DBS during melt compounding. At first, the polymer

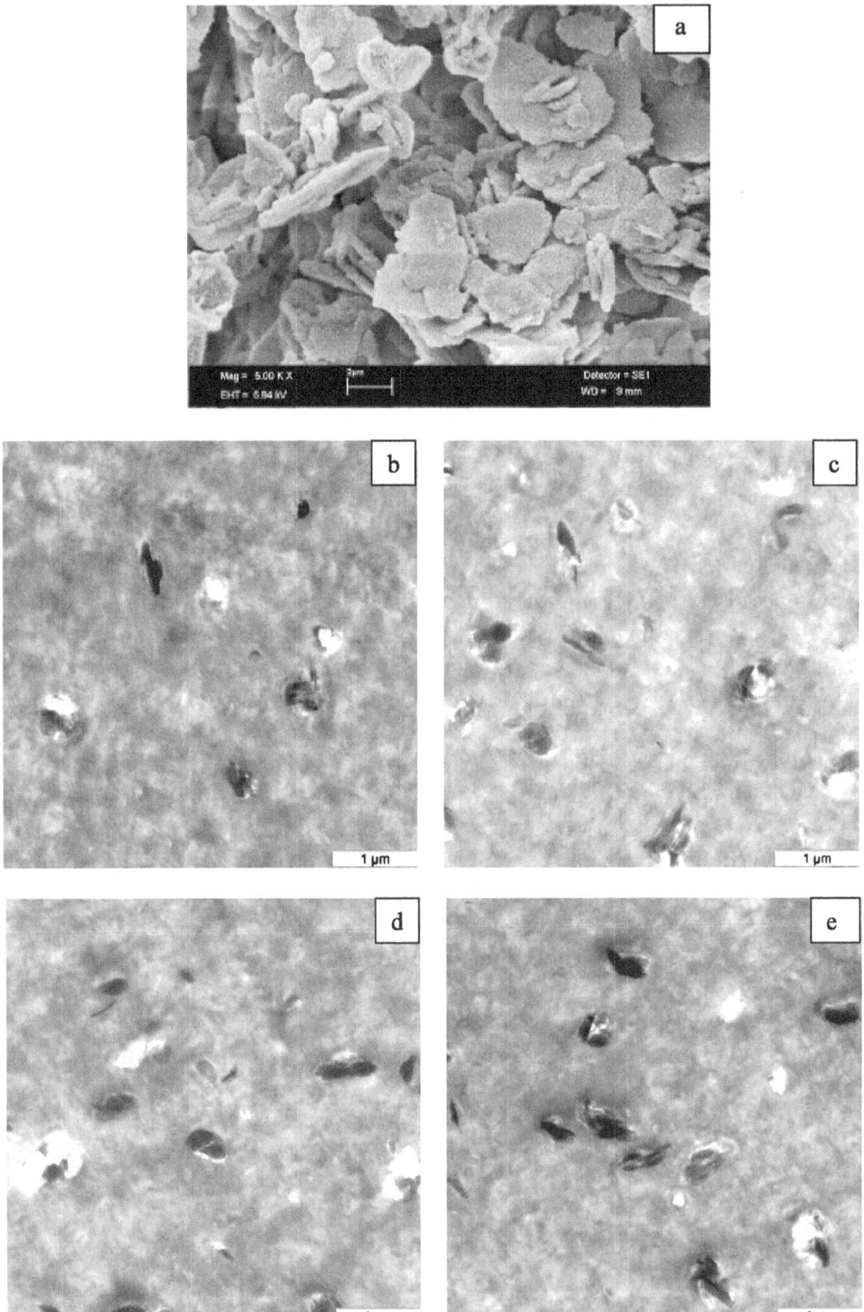

Fig. 9 Morphological analysis of the LDH particles: **a** SEM image of LDH-DBS; **b–e** low magnification TEM images of PE/LDH nanocomposites: **b** 4.72 wt %, **c** 6.89 wt %, **d** 8.95 wt %, and **e** 12.75 wt % (Reprinted from [135], with permission from Elsevier)

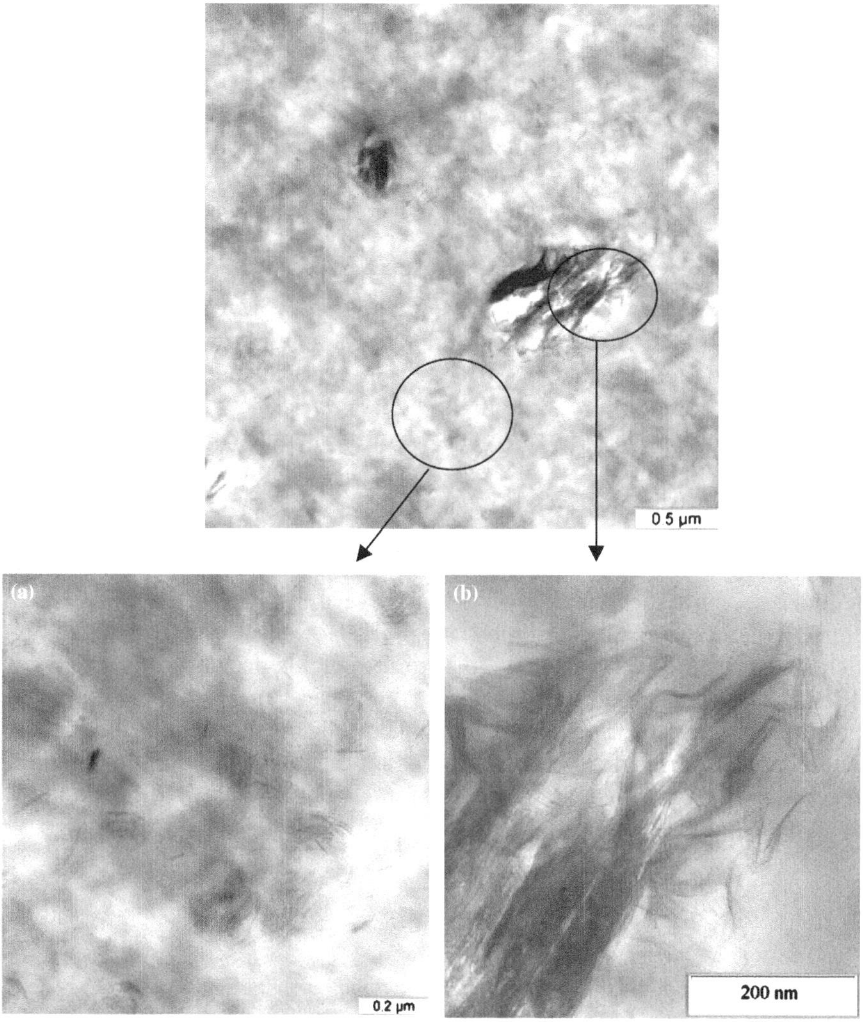

Fig. 10 TEM micrographs showing how the LDH particles are dispersed in LDPE matrix: **a** exfoliated LDH particles dispersed in the bulk matrix; **b** magnified view of the interfacial region between a primary particle and the bulk matrix showing the exfoliation process and the gathering of the exfoliated layers at the interface (Reprinted from [135], with permission from Elsevier)

chains/segments penetrate within the interlayer region of the LDH-DBS particles and push apart the metal hydroxide sheets. With time, within the mixing channel of extruder, more and more polymer chains enter the interlayer region and the shearing of the screws forces the delamination of the surface layers one by one from the surface of a large LDH-DBS particle. As a result, not only the exfoliated particle fragments are formed, but also the size of the original particles is reduced. This also explains why the average size of the primary particles observed in Fig. 9 is much smaller than that observed in the case of LDH-DBS prior to melt compounding. The breakdown of the organically modified primary LDH particles has also been observed during melt compounding in the case of polyamide 6-based nanocomposites [78]. Therefore, it is expected that conditions employed during the melt compounding (e.g., screw speed and geometry, temperature, feed rate, etc.) will have a strong influence on the extent of exfoliation of the LDH particles. Optimization of these conditions would give the best result in terms of particle dispersion [86, 87]. The delaminated crystal layers may remain either crowded in the vicinity of the parent particles, or the shearing action of extruder screw can force them to disperse throughout the matrix. This has been clearly demonstrated in Fig. 10a and b. The exfoliated layers are highly anisometric, having a very large aspect ratio and thickness much smaller than the primary LDH particles. The breakdown of the single crystal layers during the delamination process is also apparent, as the dimensions of the exfoliated particles are much lower than that of the primary particles. In addition to the shearing action during melt compounding, the low crystal layer rigidity of the LDH-type clay can be a potential reason for such breakdown (in this regard LDH has lower crystal layer rigidity than layered silicate-type clays [88]). However, these exfoliated particles do not necessarily exfoliate to a single metal hydroxide sheet, but most often to a stack of several such sheets. Further, Fig. 11 shows that the exfoliated particles are indeed dispersed to a significant extent throughout the matrix. These morphological features also prove that the melt compounding method is certainly a promising method for obtaining exfoliated particle morphology, at least partially, of LDH nanofillers in polyethylene.

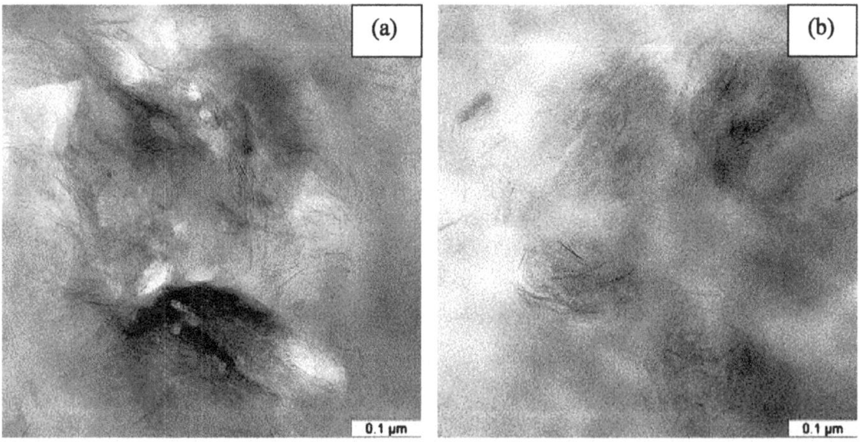

Fig. 11 High magnification TEM micrograph showing that LDH particles also undergo exfoliation into single layers: **a** PE-LDH4 and **b** PE-LDH5 (Reprinted from [135], with permission from Elsevier)

4.3 Melt Rheological Behavior

Rheological analysis of polymeric melts involves the study of the mechanical response of the melt under external stress or strain. In the case of filled polymeric systems, rheological behavior can be drastically different from that of the unfilled melts as it depends on the nature of the filler particles (structure, size, shape, surface characteristics, etc.) and the state of their dispersion in the polymer matrix [89, 90]. In fact, rheological analysis provides indirect information about the filler particle dispersion and possible particle–particle or particle–polymer interactions through measuring their influences on the flow behavior in comparison with the corresponding unfilled polymer melts. Though an indirect method, rheological analysis can be treated as complementary to the direct methods of morphological analysis, like XRD and electron microscopy, which together provide a complete picture of the state of filler particle dispersion in polymer nanocomposites. The major advantage of rheological analysis is that it provides flow behavior of the filled melt, important information in deciding suitable conditions during real melt processing of these materials.

Filled polymeric melts are viscoelastic systems and their response to shearing action depends on the ratio between the time scale of the experiment and the characteristic relaxation time of microstructures present within such systems. Here, the term microstructure means molecular entanglements in a high molecular weight polymer matrix and, additionally, a structural association of the filler particles in a filled system. When the experimental time scale is far below the longest relaxation time of the microstructure (high frequencies or shear rates), the system shows preferably the elastic response characterized by a high value of the storage modulus. In contrast, at large experimental time scales (low frequencies or shear rates) the system shows viscous response. Sufficiently high shear deformations can change and even destroy both microstructures, resulting in an entirely different material response compared to that observed when shearing does not affect the microstructures. So, the polymeric melts are characterized by a critical strain, below which the stress bears a linear relationship to the applied strain and their ratio (known as relaxation modulus) is independent of strain. The rheological behavior of polymeric melts below this critical strain is a linear viscoelastic one. Above the critical strain, due to changes in the microstructure, the relaxation modulus decreases with strain and the stress becomes a non-linear function of strain, resulting in a non-linear viscoelastic material response [91].

Therefore, a primary task before carrying intensive rheological analysis is to determine the transition point between linear and non-linear viscoelastic regimes. One simple way to determine the critical strain, or a range around it, is to subject the polymeric melt to dynamic oscillatory shear using sinusoidal

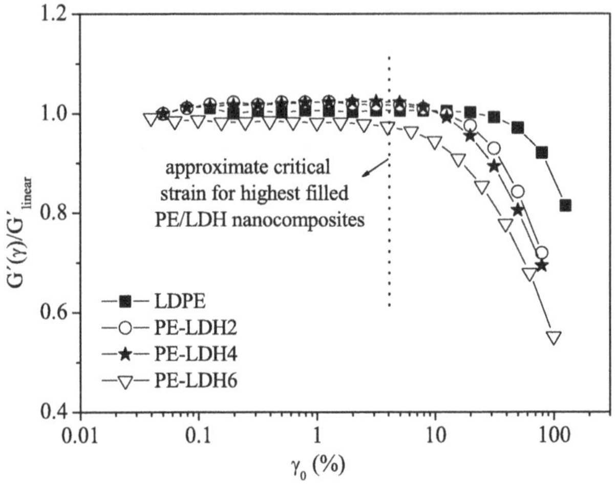

Fig. 12 Strain dependence of the storage modulus, normalized to its value in the linear regime, during the amplitude sweep experiment

strain input at constant frequency ω and strain amplitude γ_0:

$$\gamma(t) = \gamma_0 \sin \omega t \,. \tag{1}$$

The storage modulus, G', is then monitored against gradually increasing strain amplitude. The transition from a linear regime to non-linear regime is indicated in a $\log G'$ versus $\log \gamma_0$ plot by change of the storage modulus from a low strain plateau value to strain-dependent values [92]. In the case of LDPE/LDH nanocomposites, the strain amplitude sweep tests were carried out at a constant frequency of 10 rad s^{-1} and temperature of 240 °C (Fig. 12). It is apparent that unfilled polyethylene melt exhibits linear behavior up to much larger strains than filled systems, with the critical strain found within the range of 25–30% strain. In the case of LDPE/LDH nanocomposites, the critical strain decreases with increasing LDH content, presumably due to specific interactions between the filler particles as well as between the particles and polymer chains. However, for the nanocomposites with LDH concentration up to 16.20 wt %, the strain range below 5% can be taken as a safe one for studying linear viscoelastic behavior.

4.3.1 Linear Viscoelastic Behavior

The linear viscoelastic response of LDPE/LDH nanocomposites has been studied using dynamic oscillatory measurements at constant strain amplitude of 2% and frequency sweep of 0.05–100 rad s^{-1}. The response of all nanocomposites is found to be qualitatively similar in the temperature range 160–240 °C. However, the time–temperature superposition principle is not fulfilled, especially in the low frequency region (the vertical shift factors are found to change with the LDH concentration) [93]. Therefore, in the rest of this section, all experimental data will be presented at a chosen temperature of 240 °C.

In the case of unfilled polyethylene, the storage modulus is found to be lower than the loss modulus within the experimental frequency range (Fig. 13). This means the unfilled polyethylene melt has dominant viscous character in this frequency range. With the increase of frequency, the storage modulus increases faster than the loss modulus and at some frequency (which is larger than the highest value of frequency used in the present experiment) G' would cross G'', indicating increasing dominance of the elastic response. This is the typical behavior of unfilled thermoplastic melts [91]. For further analysis, $G'(\omega)$ and $G''(\omega)$ data for unfilled polyethylene have been converted into the relaxation spectrum using the commercial software "IRIS Rheology Tool Kit" [94, 95]. The longest relaxation time extracted is about 50 s at 240 °C (Table 2). It can be seen from Fig. 13 that a back conversion from the time domain to the frequency domain provides a perfect fit of the storage and loss moduli.

The LDPE/LDH nanocomposites with high LDH content exhibit a completely different behavior from that found in the case of unfilled matrix (Fig. 13). The storage modulus always remains higher than the loss modulus, indicating dominant elastic character of the material. Similar behavior is also observed in crosslinked polymers or polymer gels, where the viscous flow is

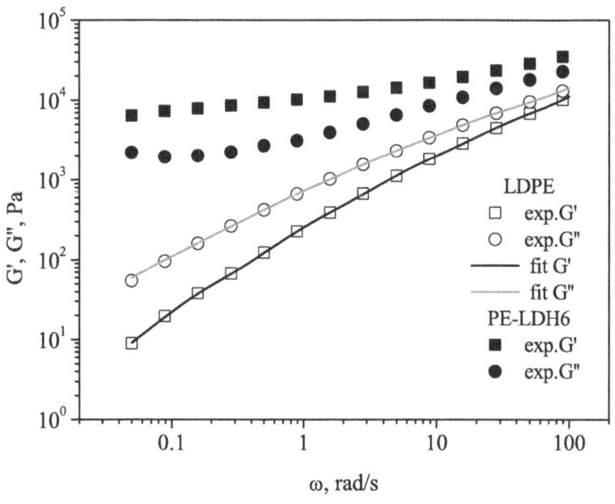

Fig. 13 Storage modulus G' and loss modulus G'' of unfilled LDPE and highly filled LDPE/LDH nanocomposite (20 wt %) in the frequency sweep experiment. Also shown are the fits of G' and G'' obtained from conversion of the LDPE relaxation spectrum (see Table 2)

Table 2 Relaxation spectrum of LDPE at 240 °C

i	g_i Pa	λ_i s
1	4.018e+04	3.553e−03
2	5.360e+03	3.256e−02
3	1.415e+03	1.830e−01
4	2.997e+02	1.032e+00
5	4.528e+01	7.033e+00
6	3.634e+00	5.146e+01

restricted by the presence of chemical constraints [96]. In the present case, the role of constraints is played by the filler particles, the high concentration of which at both nano- and microscales creates a strong physical barrier against the mobility of the polymer chains. The difference in viscoelastic response between unfilled melt and the nanocomposites strongly depends on the frequency region at which they are compared (Fig. 13). At very high frequencies, the bulky filler particles cannot fluctuate in phase with the oscillating shear flow and appear virtually static. As a result, their influence is minimized and the behavior of filled polymeric melt is solely determined by the matrix behavior [89]. At low frequencies, the influence of the filler particles may be very pronounced as the particles get enough time to follow the oscillating shear flow. Therefore, in discussing the linear viscoelastic behavior of nanocomposites the low frequency data are of prime importance. At these frequencies, the polymer chains have sufficient time to relax nearly completely, showing a low value of G'. The addition of LDH filler causes an upward shift of the low frequency values of G' so that it becomes more pronounced with increasing LDH concentration and ultimately leads to the appearance of a frequency-independent plateau (Fig. 14). This means that the system gradually develops a solid-like behavior characterized by appearance of the yield stress.

This kind of behavior can be described assuming that the total shear stress is a superposition of the stress arising in the hydrodynamically reinforced polymer matrix and the stress due to the presence of filler agglomerates [90]. Further, in the frame of the used model (for details see Sect. 4.3.2), one can show that the storage modulus, $G'_\mathrm{f} = G_\mathrm{f}/(1 + \tan^2\delta)$, and loss modulus, G''_f, of the filler structure (and hence their ratio $\tan\delta = G''_\mathrm{f}/G'_\mathrm{f}$) are independent from frequency in a wide frequency range [97]. Here G_f is the elastic (effective) modulus of the filler structure, which enters as one of the parameters into the corresponding constitutive equation for stress (see Eq. 9) [90]. Being independent of frequency, the filler contribution becomes negligible in comparison with the increasing contribution from the polymer matrix at high frequencies. Overall, the following frequency dependency of the nanocom-

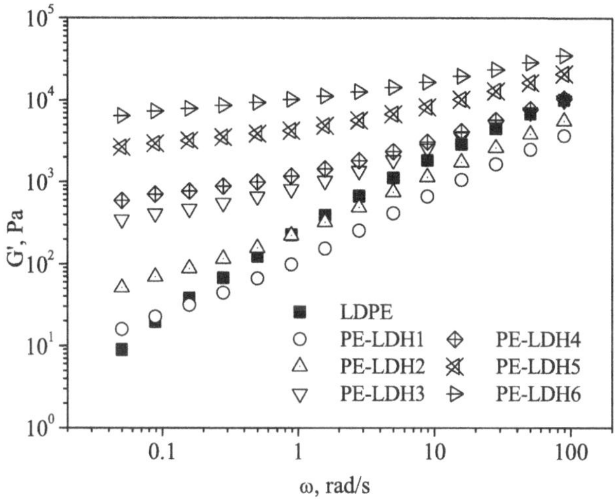

Fig. 14 Dependence of the storage modulus G' on frequency ω for unfilled LDPE and LDPE/LDH nanocomposite melts

posite storage modulus can be obtained:

$$G'(\omega) = XG'_{\mathrm{p}}(\omega) + G'_{\mathrm{f}}, \tag{2}$$

where the hydrodynamic amplification factor, X, is given by the following approximate formula [90]:

$$X = \left(1 - \frac{\phi}{\phi_{\mathrm{m}}}\right)^{-2} \approx \left(1 - \frac{\Phi}{\Phi_{\mathrm{m}}}\right)^{-2}. \tag{3}$$

Here $\phi(\Phi)$ is the volume (weight) fraction of the filler particles in the nanocomposite and ϕ_{m} (Φ_{m}) is their maximum-packing volume (weight) fraction. The volume fraction can be calculated from the weight fraction as $\phi = \Phi\rho_{\mathrm{p}}/((1-\Phi)\rho_{\mathrm{f}} + \Phi\rho_{\mathrm{p}})$ where $\rho_{\mathrm{p}} = 0.92\ \mathrm{g\,cm^{-3}}$ is the density of polymer matrix and $\rho_{\mathrm{f}} = 1.5\ \mathrm{g\,cm^{-3}}$ is the estimated density of the modified LDH particles. The latter estimation was done by taking the organic content in the modified particles to be equal to 54 wt % and the density of the unmodified LDH as $2.1\ \mathrm{g\,cm^{-3}}$.

The polymer matrix used in this study represents a blend of high molecular weight LDPE and a low molecular weight compatibilizer, a functionalized polyethylene (PE-*g*-MAH). Therefore, to determine the relaxation modulus of the polymer matrix, $G_{\mathrm{p}}(t)$, in the different nanocomposites we have used the semi-empirical "double-reptation" mixing scheme proposed for polymer blends [98]:

$$G_{\mathrm{p}}^{0.5}(t) = (1-f)G_{\mathrm{LDPE}}^{0.5}(t) + fG_{\mathrm{comp}}^{0.5}(t), \tag{4}$$

where G_{LDPE} and G_{comp} are the relaxation moduli of the LDPE and the compatibilizer, respectively, and f is the weight fraction of compatibilizer in the polymer matrix. $G_p(t)$ can be then converted into the storage modulus $G'_p(\omega)$. When LDPE is blended with the compatibilizer the storage modulus of the unfilled blend (pure matrix) is predicted to be noticeably lower than that of LDPE (Fig. 15, gray line). However, the presence of hard filler particles leads to a strong hydrodynamic reinforcement of the filled matrix, especially at high loadings. For example, the storage modulus of the filled matrix containing 16.20 wt % LDH (PE-LDH6) becomes approximately three times higher than that of LDPE in the frequency range studied (Fig. 15, black line).

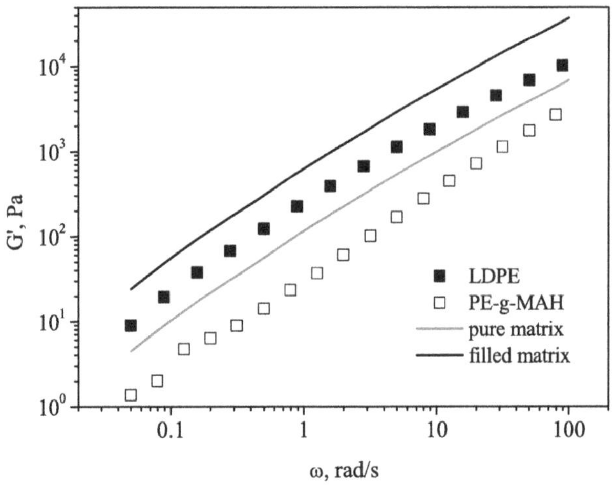

Fig. 15 Experimental data for the frequency dependence of storage modulus G' for LDPE and compatibilizer PE-g-MAH; predicted values for their blend used as a matrix in the PE-LDH6 nanocomposite: pure matrix ($X = 1$) and filled matrix ($\Phi_m = 0.35$, $X \approx 5.3$)

The effective elastic modulus of the filler structure has been determined from the low frequency data for $G'_f = G_f/(1 + \tan^2 \delta)$ and $\tan \delta = G''_f / G'_f$. With the help of Eq. 2 it is possible to reproduce reasonably well the data for highly loaded nanocomposites (>5 wt %) using the same values of $\Phi_m = 0.35$ ($\phi_m \approx 0.25$) and $\tan \delta = 0.25$ at different LDH concentrations (Fig. 16). The modulus of filler structure is found to increase dramatically with the loading: $G_f = 335$ Pa for PE-LDH3, 580 Pa for PE-LDH4, 2660 Pa for PE-LDH5, and 8000 Pa for PE-LDH6. The data for weakly loaded nanocomposites cannot be fitted using this approach, which is quite expected. First, the stress superposition assumption is only valid for highly loaded nanocomposites in which the filler agglomerate structure is already sufficiently developed [90]. Second, small deviations in the predicted storage modulus of the polymer matrix noticeably affect the estimated values of G'_f at small LDH concen-

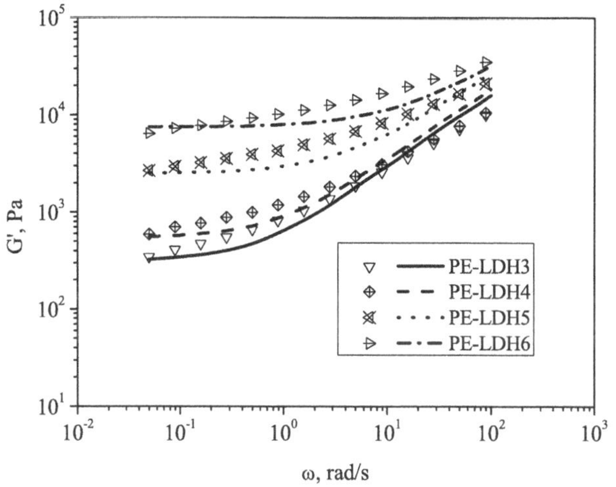

Fig. 16 Experimental (*points*) and predicted (*lines*) frequency dependence of storage modulus G' for highly filled LDPE/LDH nanocomposite melts

trations. This makes a whole fitting procedure at low filler concentration unreliable.

The strong influence of LDH concentration on the relaxation process of the nanocomposite melt can be directly observed in the stress relaxation experiment. Here, a constant strain γ_0 is applied to an initially equilibrated sample at time $t = 0$ and then the resulting shear stress, σ, is monitored as a function of time. As the applied strain is maintained at constant value, the time-dependent relaxation modulus is given by $G(t) = \sigma(t)/\gamma_0$. In the case of unfilled polyethylene, $G(t)$ decays so fast that the stress signal beyond 10 s becomes lower than the measurement range of the torque transducer of the rheometer, which causes a scattering in the $G(t)$ values (Fig. 17). This means that the unfilled melt undergoes nearly complete stress relaxation at a very short time as the longest relaxation time of LDPE does not exceed 50 s (see Table 2). The addition of a small amount of LDH (2.43 wt %, PE-LDH1) results in a significant slowing down of the decay of $G(t)$ with time. Also, with increasing LDH concentration, $G(t)$ shows a tendency to attain a non-zero residual value at long times ($t \rightarrow \infty$).

The latter effect is certainly induced by the dispersed LDH particles through particle–particle and particle–polymer interactions, which change the relaxation dynamics of the system. On the one hand, the morphological and fracture surface analysis of the LDPE/LDH nanocomposites revealed that some of polymer chains are adsorbed on the LDH particle surface and entrapped within loose particle clusters [99, 100]. Such adsorption of polymer chain segments on a rigid surface can act as an energetic barrier against the relaxation through polymer chain reptation [101]. On the other hand, the number density of exfoliated LDH layers, randomly dispersed in the ma-

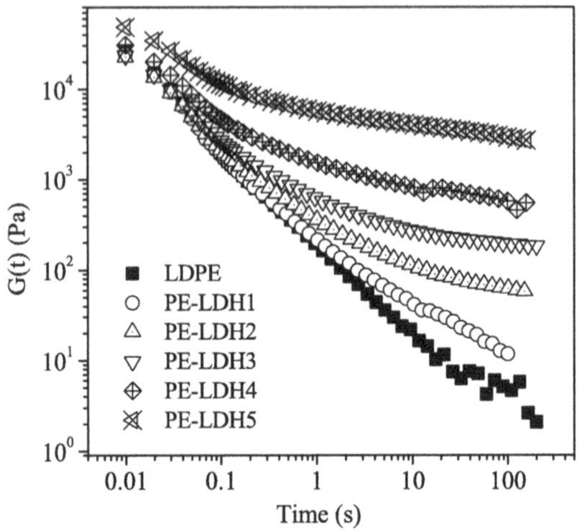

Fig. 17 Shear stress relaxation modulus for unfilled LDPE and a series of LDPE/LDH nanocomposite melts after a step strain

trix and in the vicinity of larger particle agglomerates, increases with LDH concentration, which decreases the average interparticle distance. This may lead to formation of localized network structures, in which the nanostructured particles can be oriented in some preferential direction [93, 102, 103]. Further, the close proximity and strong particle–particle interaction in highly loaded nanocomposites may cause a kind of physical jamming that will result in an extremely slow relaxation of the particle phase. In fact, even the simple model, taking into account the agglomeration of filler particles (for detail see Sect. 4.3.2), predicts the appearance of the non-zero residual stress observed experimentally (Fig. 17).

A rapid increase of the complex shear viscosity, which is especially pronounced at low frequencies, can also serve as a strong indication of the appearance of attractive filler agglomerates [104]. In the case of unfilled polyethylene and nanocomposites containing low LDH amounts, a well-defined zero-shear viscosity can be determined by extrapolating to zero frequency. However, this does not work at high LDH concentrations because the complex viscosity diverges as frequency approaches zero, indicating the presence of a positive yield stress value. Therefore, instead of zero-shear viscosities, the complex viscosities determined at a low frequency of 0.05 rad s^{-1} have been compared for different LDH concentrations (Fig. 18). A rapid (exponential) increase of the shear viscosity with the increasing LDH amount cannot be explained by taking into account only a hydrodynamic reinforcement of the polymer matrix due to the presence of filler particles. It is well known that hard particles experiencing only hydrodynamic interactions between each other, and rigid repulsions at high volume fractions, increase the shear viscosity of a suspension, η, according to the following empirical equa-

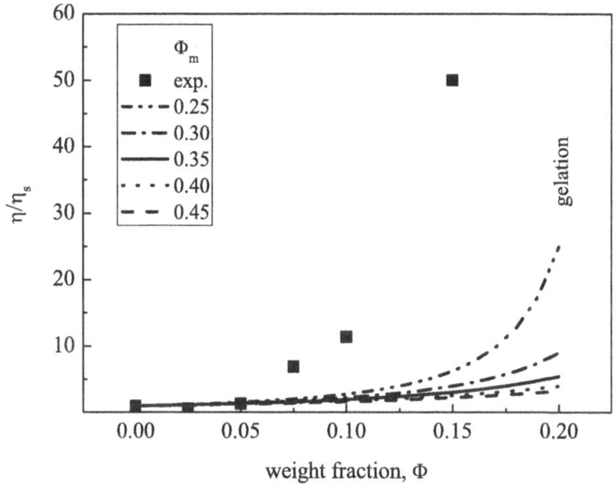

Fig. 18 Experimental (*points*) and predicted (*lines*) dependencies of dimensionless shear viscosity η/η_s on the LDH concentration at $\omega = 0.05\ \mathrm{rad\,s^{-1}}$. η_s has been chosen equal to the zero-shear viscosity of the unfilled LDPE (≈1100 Pa s)

tion valid for the particles of arbitrary shape [98]:

$$\eta = \eta_s \left(1 - \frac{\phi}{\phi_m}\right)^{-[\eta]\phi_m}, \tag{5}$$

where η_s is the shear viscosity of a host fluid (unfilled polyethylene in our case) and $[\eta]$ is the intrinsic viscosity. With the increase of the particle aspect ratio, the intrinsic viscosity increases and the maximum-packing volume fraction ϕ_m decreases, while the product $[\eta]\phi_m$ usually remains in the range 1.4–3.0. In this study, a simplified expression with $[\eta]\phi_m = 2$, which was proposed in [90] and verified in the case of filled polymer, has been used. Figure 18 shows prediction of Eq. 3 for $X = \eta/\eta_s$ at five different values of the maximum-packing weight fraction. It is clearly seen that all predicted curves lie considerably lower than that extracted from the experimental data. Only at $\Phi_m = 0.25$, the lowest value used, does Eq. 3 predict an approach to the gelation point, i.e., divergence of the shear viscosity due to crowding (jamming) of the highly anisometric hard particles. However, at $\Phi_m = 0.35$, which is the value of maximum-packing weight fraction extracted earlier from the fitting of the storage modulus, the discrepancy between predicted values and experimental data in the case of highly loaded nanocomposites is very large.

This discrepancy can be explained by a build-up of filler agglomerates that contribute considerably to the shear viscosity, especially at low frequencies. When the LDH amount exceeds 5 wt %, filled polymeric melts due to the presence of structured domains exhibit a solid-like behavior manifesting itself in a dramatic increase of the shear viscosity and elastic modulus. During low-strain and low-frequency oscillatory shearing, the matrix mobility is severely restricted by such particulate domains. Additionally, the polymer chains, which are entrapped between particle clusters or constrained by the clay particles through intercalation and adsorption, experience larger effective strain compared to the unconfined chains [105]. This can lead to enhanced shearing behavior of the nanocomposite melts even at the low shear rates arising in low-frequency oscillatory measurements [106, 107].

4.3.2 Non-linear Viscoelastic Behavior

In the linear viscoelastic regime, when the filled polymeric melts are sheared at small strain or strain rates, various microstructures present in the system are not destroyed, but rather undergo a slight reorganization. However, in most practical situations these materials experience a much severer shearing at high strain or strain rates (within the non-linear viscoelastic regime). For example, during extrusion the rotational motion of a screw generates extremely high shear rates ($\sim 1-10\,s^{-1}$ in the laboratory extruders and $\sim 10^2-10^4\,s^{-1}$ in the industrial extruders). Such strong shearing is often necessary to facilitate breakdown of large filler particles for achieving a better dispersion in the filled polymer nanocomposites. Already, shearing of unfilled polymeric melts in the non-linear regime causes noticeable changes in the internal structure of the system [98]. These changes become much more pronounced in the case of a filled system, especially when the filler particles show strong attractive interactions between each other. Such filled polymeric melts usually exhibit a so-called thixotropic behavior. This means that during shearing in the non-linear regime, the filler agglomerates present in the system suffer a structural breakdown whose extent depends on the magnitude and duration of the shear force, and when shearing is stopped the regeneration of agglomerates takes place [89]. The breakdown and regeneration of the filler microstructure, both being time-dependent processes, introduce additional time constants in the response behavior of the filled polymeric melt under non-linear shearing. To separate this effect from the non-linear response of the polyethylene matrix, first some studies of the unfilled LDPE have been carried out.

The LDPE used in the present investigation is a high molecular weight commercial polymer. This type of LDPE is characterized by a long-chain branching with multiple branch points on a single polymer chain. The multiple long-chain branching, in addition to the chain entanglements, causes a high steric hindrance against the process of chain stretching and orientation under the influence of shear flow and thus strongly affects a non-linear response at high shear rates. At low shear rates ($\dot{\gamma} < 0.03\,s^{-1}$), the stress response of unfilled polyethylene is characterized by a monotonous increase with time until a steady state is reached (Fig. 19). With increasing shear rate,

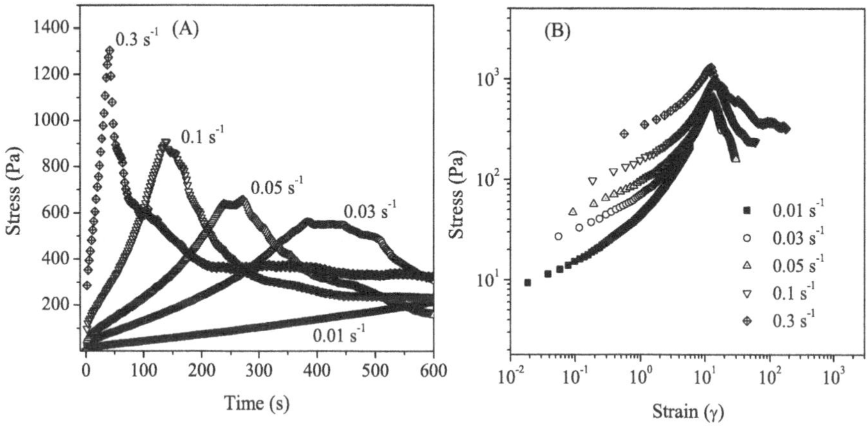

Fig. 19 Onset of steady shear flow for unfilled LDPE melt at different shear rates represented as **a** a time dependence and **b** a corresponding strain dependence. $T = 240\,°\mathrm{C}$

a tendency to form a stress maximum before attaining the steady state is observed. With further increase in shear rate, this stress maximum becomes more pronounced and its position shifts to the left on the time axis. Thus, a distinct stress overshoot can be observed at the start of shear flow at high shear rates. Such non-linear flow behavior is well known and is theoretically described in the case of high molecular weight polymer melts and concentrated polymer solutions [108–110].

In the case of unfilled polyethylene melt, the shear strain $\gamma_{\mathrm{max}} = \dot{\gamma} t_{\mathrm{max}}$, at which the stress overshoot is observed, is independent of the shear rate within the experimental range (Fig. 19b). The Doi–Edwards theory proposed for monodisperse linear polymer melts predicts for γ_{max} a value of about 2, roughly independent of the shear rate [109, 111]. However, in the present case, γ_{max} is found to be about 20, which cannot be explained in the frame of Doi–Edwards theory. Similar deviation has already been reported for commercial LDPEs (e.g., $\gamma_{\mathrm{max}} = 7$) by Wagner [112] and for other high molecular weight branched polymers by Osaki [113]. Such deviation is attributed not to the sample polydispersity that mostly influence the linear properties (i.e., the stress relaxation modulus and the complex dynamic modulus) but rather to a long-chain branching associated with commercial low-density polyethylene melts [113, 114]. Irregularly spaced long side branches entangled with surrounding chains not only suppress reptation of polymer strands but also suppress their retraction after a step strain [98]. The effects of long side branching on rheological properties are difficult to consider theoretically. Here, following [97], we simply note that the presence of branch points at both ends of a polymer strand leads to much smaller shear thinning, i.e., higher values of γ_{max}. This kind of behavior can be easily understood if one takes into account that increasing the number of branch points shifts the behavior of the system to that of a polymer network in which there is no retraction of strands and thus no shear thinning effects.

The onset of steady shear flow for unfilled LDPE melt and a series of highly filled LDPE/LDH nanocomposites has been compared at $\dot{\gamma} = 0.3\ \mathrm{s}^{-1}$ (Fig. 20). While the unfilled polymer melt exhibits a single stress overshoot due to the presence of entanglements and long-chain branching, the non-linear response of nanocomposite systems is characterized by two stress overshoots. The first overshoot appears at γ_{max} of about 2.2 and, most likely, is due to the presence of filler agglomerates, as a similar value of γ_{max} has been reported for the filler structure in the case of silicate–polypropylene nanocomposites [115]. The position of the second overshoot strongly depends on the filler concentration. In the PE-LDH2 nanocomposite (5 wt %) it is found at shear strain of about 15, close to that exhibited by the branched polymer matrix ($\gamma_{max} \approx 20$), and then shifts with the increase of LDH loading to smaller deformations, eventually merging with the first overshoot.

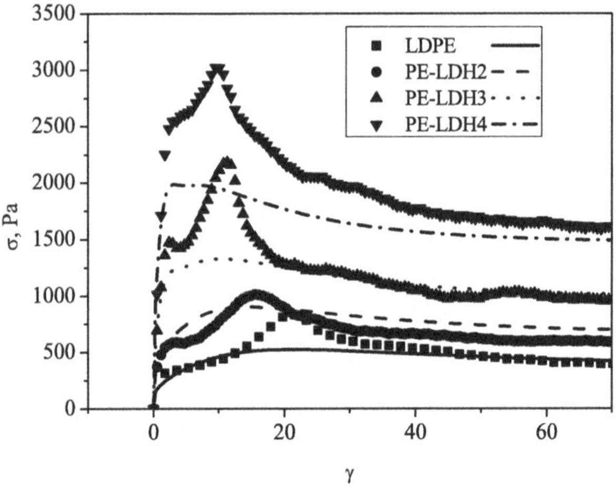

Fig. 20 Onset of steady shear flow for the unfilled LDPE melt and a series of highly filled LDPE/LDH nanocomposites at $\dot{\gamma}_0 = 0.3\ \mathrm{s}^{-1}$: experimental data (*points*) and predicted results (*lines*)

It is generally accepted that appearance of stress overshoot at the beginning of steady shear is related to the accumulation of stress in the particle phase and its subsequent release due to the rupture of various particulate structures [90, 97, 106, 116]. The particle phase mainly consists of physically associated particle domains such as the soft particle clusters and localized network structures formed both by primary clay particles and the nanoscopic exfoliated clay layers [106]. When the nanocomposite melt is subjected to shearing, both the particle phase and the matrix phase respond according to their characteristic structural rigidities. The elastic modulus of the particulate structure is certainly much higher than that of the polymer matrix. This is because the force that facilitates the formation of various particulate structures in the melt state is usually of electrostatic nature and also coupled with thermodynamic incompatibility between the particle and the polymer phases. In the case of LDH particles, this force is expected to be very strong because of the high surface charge density. The elasticity of the polymer matrix is considerably lower, being mostly related to the molecular entanglement and branching, i.e., to the presence of topological constraints. Therefore, at the inception of shearing the phase with higher elastic modulus (i.e., the particle phase) responds first. The rupturing process continues till the steady state is reached, when the primary particles become orientated with respect to the flow direction [106].

To gain a deeper insight in the physics behind this complex response, we shall try to provide a quantitative description, which is mostly based on the guidelines proposed in [87] and [94]. In this description the influence of particle orientation on the flow behavior will be completely neglected, although some authors try to reproduce the stress overshoot using a purely orientational model for the particle phase [115]. Generally, the total shear stress, σ, in the highly filled polymer melt can be represented as a superposition of the stress, σ_{p}, arising in the hydrodynamically reinforced polymer matrix and the stress, σ_{f}, due to the filler structure [90]:

$$\sigma = \sigma_{\mathrm{p}} + \sigma_{\mathrm{f}} \,. \tag{6}$$

To describe the polymer stress in this equation, one can probe any of rheological constitutive models proposed for the long-chain branched polymers: the partially extending convection (PEC) model of R. Larson, [117], the molecular stress function (MSF) theory of M. Wagner et al. [118, 119], the modified extended pom-pom (mXPP) model of M.H. Verbeeten et al. [120], etc. Here, the PEC model has been chosen as it can be easily tuned to describe the overshoot position for a wide class of polymers by changing a value of the non-linear parameter β. Thus, $\beta = 1$ in the case of linear polymer (Doi–Edwards limit), $0 < \beta < 1$ in the case of branched polymers, and $\beta = 0$ in the case of networks (affine deformation). The polymer stress in the PEC model with one relaxation mode is described as follows [117]:

$$(\sigma_{\mathrm{p}})_{(1)} + \frac{\beta}{3G_{\mathrm{p}}}(\dot{\gamma} : \sigma_{\mathrm{p}})(\sigma_{\mathrm{p}} + G_{\mathrm{p}}\delta) + \lambda_{\mathrm{p}}^{-1}\sigma_{\mathrm{p}} = G_{\mathrm{p}}\dot{\gamma} \,, \tag{7}$$

where $(\boldsymbol{A})_{(1)}$ denotes the upper-convected time derivative of a tensor $\boldsymbol{A}$ [121], $\dot{\gamma}$ is the deformation tensor, and δ is the unit tensor. Further, the following assumptions are made about the modulus, G_{p}, and the relaxation time, λ_{p}, of the hydrodynamically reinforced polymer matrix [90]:

$$G_{\mathrm{p}} = G_{\mathrm{p}}^{0} X(\Phi) \quad \text{and} \quad \lambda_{\mathrm{p}} = \lambda_{\mathrm{p}}^{0} \,, \tag{8}$$

where G_p^0 and λ_p^0 are the elastic modulus and the relaxation time of the unfilled polymer matrix, respectively, and the hydrodynamic amplification factor X is given by Eq. 3.

The viscoelastic stress experienced by the filler structure is described by a modified Maxwell model, i.e., the expression used for a modified Maxwell element. In this case it is assumed that elastic deformations in the filler phase are so small that a linear approximation can be used [90]:

$$(\sigma_f)_{(1)} + \frac{G_f}{\eta(t)}\sigma_f = G_f\dot{\gamma} \,. \tag{9}$$

Here G_f is the elastic (effective) modulus of the filler structure and η is its time-dependent viscosity. The latter should diverge at zero time to describe a solid-like behavior in the equilibrium state and attain at $t \to \infty$ a finite steady value depending on the shear rate. To find an expression for $\eta(t)$, it is assumed that the filler structure is stabilized by a number of bonds that can be broken by the shearing in a non-linear regime and can be rebuilt again during the quiescent time. The total number of bonds (broken and unbroken) does not change. Hence, a usual kinetic equation can be written for the change in the fraction of broken bonds ξ_b:

$$\frac{d}{dt}\xi_b = a\,|\dot{\gamma}_0|\,(1-\xi_b) - b\xi_b \,, \tag{10}$$

where the first term on the right side describes the bond rupture in a flow with the shear rate $\dot{\gamma}_0$, while the second term allows the bond rebuilding driven by strong attractive interactions between the particles. Here a is the rupture strength and b is a positive constant whose meaning will become clear later. On the start-up of shear flow, the fraction of broken bonds increases according to:

$$\xi_b(t) = \frac{a\dot{\gamma}_0}{a\dot{\gamma}_0 + b}\left(1 - e^{-t/\lambda_\gamma}\right) \quad \text{with} \quad \lambda_\gamma = \frac{1}{a\dot{\gamma}_0 + b} \,. \tag{11}$$

During relaxation after the flow is switched off:

$$\xi_b(t) = \frac{a\dot{\gamma}_0}{a\dot{\gamma}_0 + b}\, e^{-t/\lambda_0} \quad \text{with} \quad \lambda_0 = 1/b \,. \tag{12}$$

Thus, the constants a and b define the relaxation times in the presence (λ_γ) and in the absence(λ_0) of flow. In the non-linear regime, i.e., at high shear rates, λ_γ may become significantly lower than λ_0, which results in two totally different time scales in the system [122].

The time-dependent viscosity function can be now defined as:

$$\eta(t) = \frac{\eta_0}{\xi_b(t)} \,. \tag{13}$$

In the equilibrium state ($t = 0$), the number of broken bonds is equal to zero, and hence $\eta(0) \to \infty$. On the start-up of shear flow, the shear stress is given by:

$$\sigma_f(t) = G_f\dot{\gamma}_0 \int_0^t \exp\left\{-\frac{a\dot{\gamma}_0\lambda_\gamma}{\lambda_\eta}\left[(t-t') + \lambda_\gamma\left(e^{-t/\lambda_\gamma} - e^{-t'/\lambda_\gamma}\right)\right]\right\} dt' . \quad (14)$$

It goes through a maximum, the strength of which increases with increase of $\lambda_\eta = \eta_0/G_f$ and attains a steady value:

$$\sigma_f^{st} = \frac{G_f\lambda_\eta}{a\lambda_\gamma} = \frac{G_f\lambda_\eta}{a\lambda_0}(1 + a\dot{\gamma}_0\lambda_0) . \quad (15)$$

Thus, the modified Maxwell equation predicts the appearance of the yield stress observable in highly filled nanocomposites:

$$\sigma_f^Y = \frac{\eta_0}{a\lambda_0} . \quad (16)$$

Also, this model is able to predict the appearance of residual shear stress observed in the stress relaxation experiments after a sudden shearing displacement γ_0 (Fig. 17). According to the model, the shear stress experienced by the filler structure decreases with time as:

$$\sigma_f(t) = G_f\gamma_0 \exp\left[-\frac{a\gamma_0\lambda_0}{\lambda_\eta}\left(1 - e^{-t/\lambda_0}\right)\right] , \quad (17)$$

and thus the stress relaxation modulus never relaxes to a zero value:

$$G(t \to \infty) = G_f \exp\left[-a\gamma_0\frac{\lambda_0}{\lambda_\eta}\right] , \quad (18)$$

in accordance with experimental data shown earlier (Fig. 17).

To reproduce the complex response at a start up of shear flow for a series of the LDPE/LDH nanocomposites (Fig. 20), it is necessary to take into account the shift of the second stress overshoot to smaller deformations with increasing LDH loading. To our knowledge, this shift can be explained by the effect of strain amplification in the polymer matrix found previously in the case of filled elastomers [103]. Upon shearing, the hard filler particles cannot be stretched; however, they can reorganize their positions in the polymer matrix, which hence experiences a noticeably higher effective deformation, γ_{eff}, than the strain externally applied to the sheared sample, γ_0 [103]:

$$\gamma_{eff} = \gamma_0 X(\Phi) , \quad (19)$$

where X is again the hydrodynamic amplification factor given by Eq. 3. Hence, the polymer matrix is sheared with an effective shear rate given by:

$$\dot{\gamma}_{eff} = \dot{\gamma}_0 X(\Phi) . \quad (20)$$

This final assumption allows extraction of a value of the maximum weight-packing fraction Φ_m from the dependence of overshoot position on the particle concentration (Table 3). It is found that $\Phi_m = 0.3$, which is slightly less

Table 3 Experimentally measured (γ_{max}^{exp}) and theoretically predicted (γ_{max}^{th}) positions of the stress overshoot as well as effective filler modulus (G_f) as functions of weight fraction

Φ	0.0	0.05	0.075	0.10
γ_{max}^{exp}	21.6 ± 0.9	15.6 ± 0.3	11.4 ± 0.3	9.6 ± 0.3
γ_{max}^{th}	20.5	14.3	11.7	9.4
G_f, Pa	0	40	335	580

than the value extracted from the fitting of the frequency dependencies of the storage modulus (see Sect. 4.3.1). If strain amplification in the polymer matrix is caused not only by hydrodynamic reinforcement but also by absorbance of polymer chains on the particle surface, it will explain a somewhat lower value of Φ_m.

Figure 20 shows a comparison of the experimentally measured shear stress and that predicted using Eqs. 6–9 at $\dot{\gamma}_0 = 0.3\ s^{-1}$. In the case of unfilled polyethylene, the stress overshoot is a rather broad one and its maximum is found at a strain of about 20 (see Table 3). This is quite a high value if one compares it with much lower literature values of about 10 [109]. Partially, broadening of the peak and its shift to the higher values of strain can be attributed to the use of the plate–plate geometry in which the shear rate is not constant but grows with the distance from the rotational axis, r, as:

$$\dot{\gamma}(r) = \frac{r}{R}\dot{\gamma}_0 , \tag{21}$$

where R is the plate radius and $\dot{\gamma}_0$ is the shear rate at the edge of the plates. While both the position and strength of the shear overshoot strongly depend on the applied shear rate, the effect of varying shear rate should be properly accounted for. In computation, the circle plate was divided into ten equal circular segments, for each of which the partial stress was calculated taking an average shear rate in the segment. The summation was made over all partial stresses to calculate the total stress. This procedure was used for the polymer phase as well as for the filler phase. Further, it is worth noting that the relaxation spectrum extracted for LDPE from the linear data (Table 2) is not sufficient to describe its non-linear response. To predict the appearance of stress overshoot in the frame of the PEC model, one has to use additionally a long-time relaxation mode ($g_7 = 30$ Pa, $\lambda_7 = 1000$ s, $\beta = 0.012$). Still, the strength of overshoot remains considerably underestimated (using the commercial software IRIS, it has been checked that the MSF theory is also unable to reproduce such pronounced overshoot at a strain of about 20).

To calculate a stress contribution from the agglomerated filler structure, the values of filler elastic modulus extracted from the fitting of the storage and loss moduli have been used (Table 3). Two relaxation times characterizing the filler structure λ_0 and λ_η and the rupture strength a have been chosen in such a way that, on the one hand, the filler overshoot appears at a strain of 2.2, i.e., the value found experimentally. On the other hand, their combination has to give a value of $\tan\delta = 2a\gamma_0\lambda_0/(\pi\lambda_\eta)$ extracted from the fitting of complex modulus: $\tan\delta = 0.25$ at $\gamma_0 = 0.02$. The best set of parameters is $\lambda_0 = 100$ s, $a = 0.1$ ($\lambda_\gamma = 25$ s at $\dot{\gamma}_0 = 0.3\ \mathrm{s}^{-1}$), and $\lambda_\eta = 0.5$. If one compares experimental data and predicted values for the time-dependent shear stress in the case of nanocomposite melts (Fig. 20), it can be seen that the superposition approach with the parameters chosen describes relatively well all characteristic features observed: the rapid increase of steady-state shear stress with the LDH loading, the shift of the second overshoot for highly loaded nanocomposites, and the presence of the first overshoot due to the rupture of filler structure under strong shearing. However, the predicted overshoot is only visible in the case of the highly loaded nanocomposite (PE-LDH4 in Fig. 20) being strongly smeared at lower weight fractions in comparison with the experimentally measured overshoot. This discrepancy is mainly due to insufficient reproduction of the LDPE overshoot. Overall, the quantitative description proposed in this study, utilizing the ideas of structural breakdown of the filler agglomerates during non-linear shearing and hydrodynamic reinforcement of the polymer matrix, works reasonably well. One should not forget that even prediction of the non-linear behavior of long-side branched polymers remains a very challenging problem.

The dynamic oscillatory shearing at low constant frequency showed previously that at low strain amplitude the storage modulus of the nanocomposite melts remains constant and independent of applied strain (Fig. 12), whereas above a certain critical strain, dependent on the LDH concentration, G' decreases with increasing strain. Therefore, kinetics of the structural breakdown of the filler agglomerates and their regeneration process can be investigated by subjecting a filled melt to two subsequent oscillatory shearing steps at constant frequency. The first step should be made at high strain (in the non-linear regime) to study evolution of the structural breakdown with time, and the second step at low strain (in the linear regime) to study regeneration of the structure. It is meaningful to characterize each step by its effective shear rate:

$$\dot{\gamma}_{\mathrm{eff}} = \frac{2}{\pi}\gamma_0\omega \tag{22}$$

calculated as an average over half period of the instantaneous shear rate $\dot{\gamma} = \gamma_0\omega\cos\omega t$. In this study, $\gamma_0 = 0.5$ has been chosen for the first step (at this strain the unfilled polyethylene does not yet show a strain-dependent amplitude) and $\gamma_0 = 0.02$ for the second step, while frequency was kept at 1 rad s^{-1}. This gives $\dot{\gamma}_{\mathrm{eff}} \approx 0.32\ \mathrm{s}^{-1}$ for the first step and $\dot{\gamma}_{\mathrm{eff}} \approx 0.01\ \mathrm{s}^{-1}$ for the second step.

It can be seen that prolonged oscillatory shearing at $\gamma_0 = 0.5$ results in the decrease of storage modulus, i.e., the nanocomposite melts gradually lose their elastic character (Fig. 21A). Initially, the storage modulus decreases fast, but later, upon approaching a steady state, the decrease slows down. This kind of behavior can be understood if one considers that the rate of structural breakdown depends not only on the effective shear rate but also on the size of particulate agglomerates (see Eq. 10) [123]. At the beginning of oscillatory shearing, the nanocomposite melt is in a state that is characterized by the presence of relatively large aggregated structures and structured domains, and thus the breakage rate is fast. As the shearing continues the average size of agglomerates is progressively reduced and, hence, the rate of structural

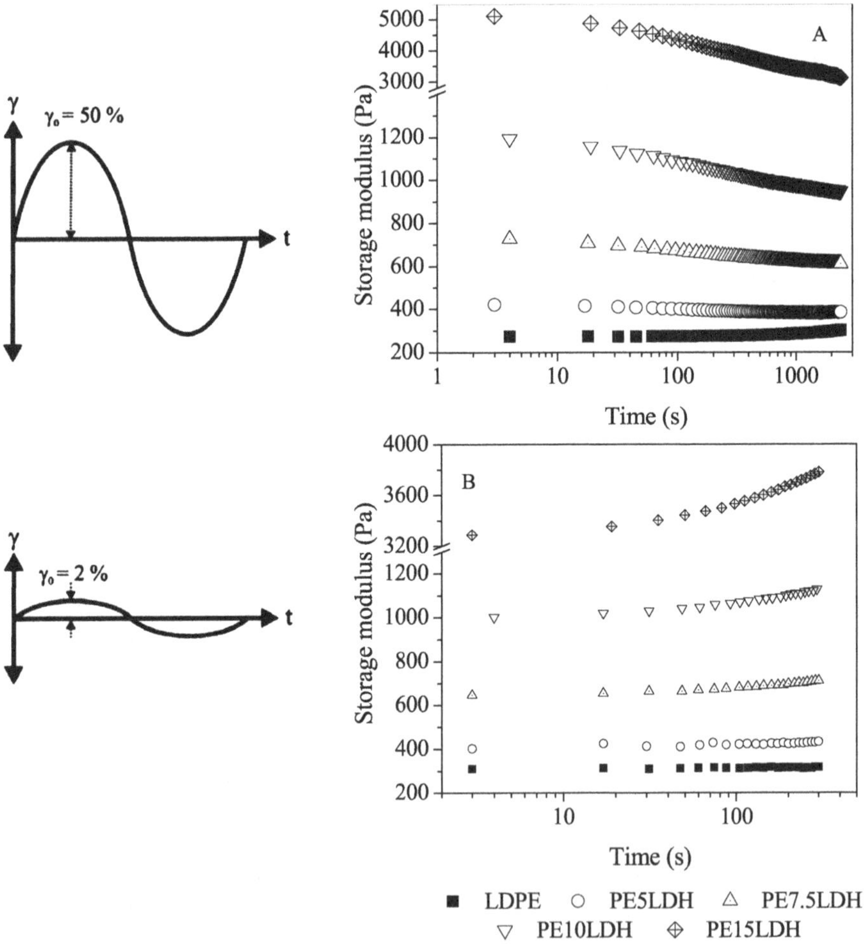

Fig. 21 Variation of storage modulus with time during oscillatory shearing **A** in the non-linear regime ($\gamma_0 = 0.5$) and **B** in the linear regime ($\gamma_0 = 0.02$). $\omega = 1\ \mathrm{rad\,s^{-1}}$

breakdown slows down. When the steady state is reached, equilibrium between the processes of structural breakdown and regeneration is achieved. At this stage, the size of the particle aggregates and the structured domains is determined solely by the applied shear rate. The rate of structural breakdown strongly depends on LDH concentration, as the latter determines both the extent and size of the structural association among the dispersed particles in the nanocomposites. The higher the concentration, the larger is the size and number density of particle aggregates and structured domains, and the faster is the breakage rate. The changes in the case of unfilled polyethylene are insignificant in comparison with the nanocomposite melts. The storage modulus even slightly increases after prolonged period of oscillatory shearing. This can be due to gradual increase of the molecular weight of LDPE held for a long time in the rheometer chamber at elevated temperature of 240 °C. For example, a noticeable increase of the molecular weight has been observed after extrusion of LDPE in a twin-screw extruder at 200 °C.

The effect of structural regeneration on the storage modulus of the nanocomposite melts is shown in Fig. 21B. Since the shearing at low effective shear rate does not destroy the microstructures within a polymeric melt, the second shearing step can be treated as equivalent to the rest period. This means that the diffusion kinetics of dispersed LDH particles is not affected by this small strain oscillatory shearing. Thus, during the second step the process of structural breakdown is stopped and the separated and oriented particles begin now to build the agglomerated structures similar to those found at the zero-shear equilibrium state. Simple Brownian motion cannot be responsible for this regeneration process as diffusion of non-interacting filler particles is estimated to be too slow to be taken into account in the explanation of this effect [124]. Rather, strong attractive interactions between the dispersed clay particles, and thermodynamic incompatibility between the particle and the polymer phases, facilitate such fast diffusion of the clay particles within a highly viscous medium [106]. The structural regeneration causes an exponential increase in the storage modulus of the melt. However, in contrast to the first step, a steady state is not achieved within the experimental time scale. This observation reflects the presence of two totally different time scales in the system: $\lambda_0 \sim 10^3$ s in the absence of shear flow or at very low $\dot{\gamma}_{\mathrm{eff}}$, and $\lambda_\gamma \ll \lambda_0$ at high effective shear rates. This is a characteristic feature of the thixotropic response. Further, it has been found that the rate of recovery increases with the LDH concentration. This effect is presumably due to a smaller average separation between the particles at the end of structural breakdown at higher concentrations, which results in a relatively shorter diffusion path length during the regeneration process.

To conclude this section, the LDPE/LDH nanocomposites, like many polyolefin/layered silicate-based nanocomposites, represent a highly inhomogeneous system with almost unknown interactions between the LDH particles and the particles and polymer matrix. Therefore, the rheological analysis carried out in this study has not been intended to provide a microscopic description of the structural evolution in the PE-LDH nanocomposites, but rather to extract some regularity. In general, the structural behavior of the PE-LDH nanocomposites under shear flow is similar to the behavior of filled elastomers, for which breakdown of filler clusters at increasing strain and their re-aggregation at decreasing strain was observed under oscillatory shear (Payne effect [103]). Further, similar to the filled elastomers, we found considerable strain amplification in the polymer matrix caused by the presence of hard particle inclusions.

4.4 Thermal Properties

Thermogravimetric (TGA) analysis of the LDPE/LDH nanocomposites prepared by the melt compounding method is shown in Fig. 22. The thermal decomposition of pure polyethylene is characterized by two major temperature regions of weight loss: first a sharp loss in the temperature range 335–410 °C (about 70.0% weight loss takes place in this region) and, second, a relatively weak loss in the temperature range 410–475 °C (more than 25.0% weight loss takes place) (Fig. 22a). In the differential thermogravimetric (DTG) plot, Fig. 22b, these two regions appear as a sharp decomposition peak at about 375 °C and a broad peak, respectively. The presence of LDH causes distinct changes in the thermal decomposition behavior in comparison to the unfilled polyethylene. With the addition of only 2.43 wt % LDH, the first decomposition stage in unfilled polyethylene is not only shifted to a higher temperature range (the corresponding decomposition peak in DTG shifts from about 375 °C to about 395 °C), but also the extent of weight loss during this stage decreases from about 70 wt % to about 25 wt % (Fig. 22a). With the further increase in LDH concentration, the low temperature decomposition peak is completely suppressed and most of the material is decomposed in the higher temperature range starting above 400 °C. Further, the broad decomposition peak at the higher temperature range observed in unfilled polyethylene becomes progressively sharp and is divided into two decomposition maxima around 430 °C and 460 °C in the nanocomposites containing high loading of LDH (Fig. 22b). Usually, the comparison of thermal stability for polymeric materials using TGA plots is carried out in terms of two temperatures. One is the onset of decomposition temperature ($T_{0.10}$), i.e., the temperature at which 10.0% weight loss takes place, and the other is the decomposition temperature ($T_{0.50}$), i.e., the temperature at which 50.0% weight loss takes place. Figure 23a shows the comparison of these two temperatures for unfilled polyethylene and the nanocomposites. The onset of decomposition is delayed significantly by the addition of 2.43 wt % LDH. In fact, with increasing LDH concentration, the onset decomposition temperature $T_{0.10}$ increases first and then stabilizes after about 5–6 wt % LDH

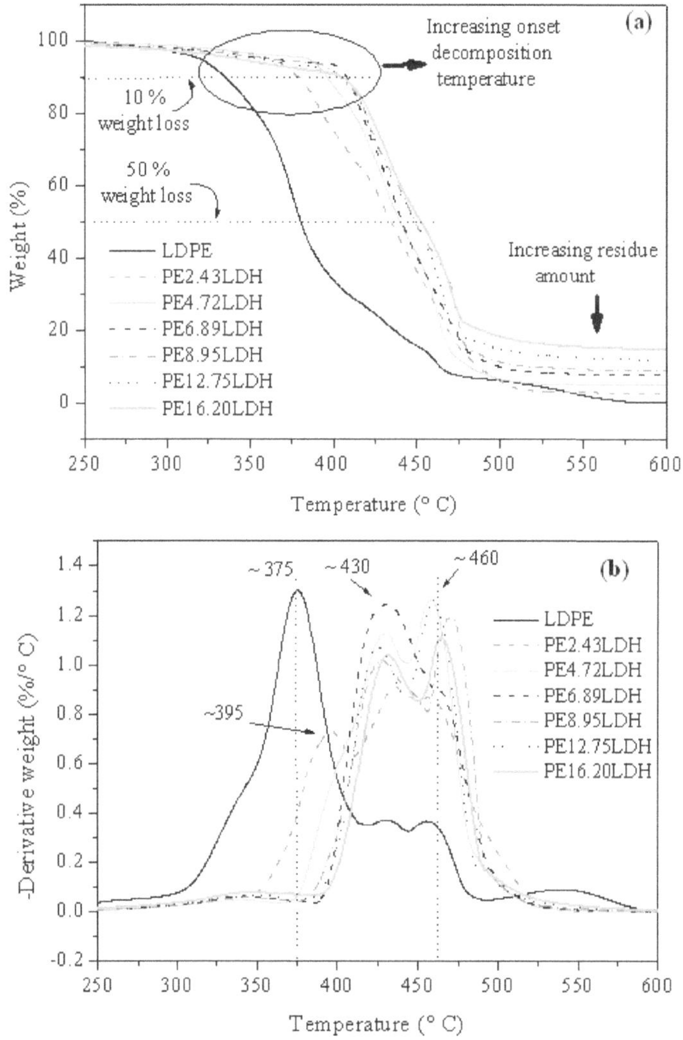

Fig. 22 TGA plots of polyethylene/LDH nanocomposites prepared by melt-mixing in extruder: **a** TGA plot, **b** DTG plot. *PEXLDH* is the sample designation, where *X* represents the weight percent of LDH) (Reprinted from [135], with permission from Elsevier)

concentration. The decomposition temperature, $T_{0.50}$ also increases with increasing LDH concentration. With about 9 wt % LDH concentration, $T_{0.50}$ increases by nearly 70 °C, as shown in Fig. 23a.

The char yield after combustion in air is nearly zero for unfilled polyethylene as its organic content (hydrocarbon) is completely converted into gaseous products. We assume that the polymeric material and the organic part in LDH-DBS are completely burnt and lost as gaseous products after

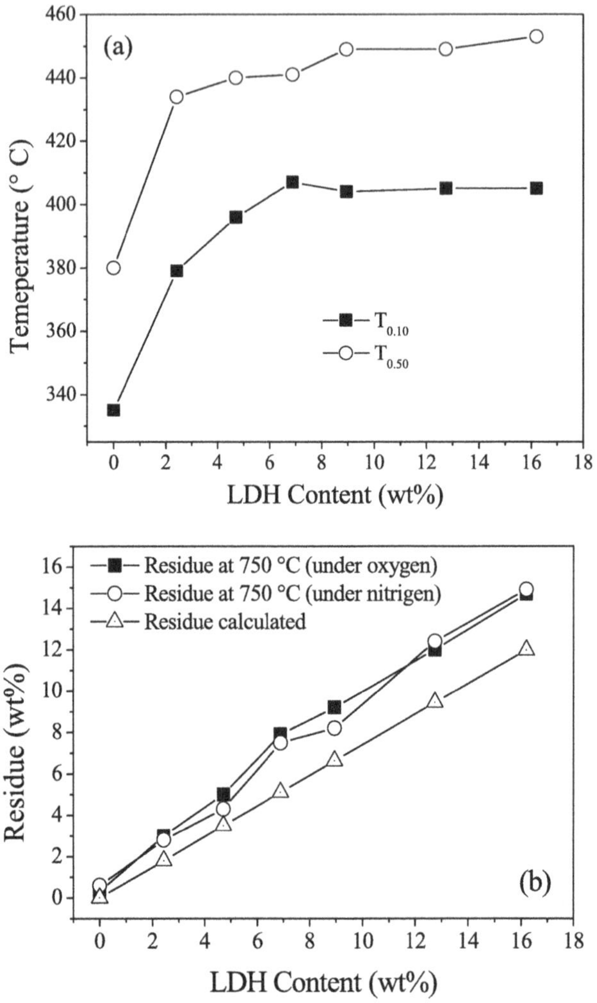

Fig. 23 The influence of LDH concentration on **a** the decomposition temperatures at 10% weight loss ($T_{0.10}$) and 50% weight loss ($T_{0.50}$), and **b** the amount of residue at 750 °C and its comparison with the theoretically calculated value (Reprinted from [135], with permission from Elsevier)

heating up to 750 °C. The amount of residue left after heating can be calculated from the approximate formula of LDH-DBS. This means that the residue formed after burning of the composite comes from the LDH and consists of the metal oxides only. Calculation of the combustion residue is based on the approximate composition of LDH-DBS, which contains about 46.0% of its total weight as metal hydroxide. The calculated value and the experimentally obtained value from TGA are compared in Fig. 23b. It is apparent that the experimentally obtained values for the nanocomposites are significantly higher than the calculated value. Though the calculation was based on an approximate formula of LDH-DBS, the difference is quite significant and cannot be taken as simple calculation error. Again, from the appearance of the residue of these composites obtained during limited oxygen index (LOI) or UL94 testing, it can be said that the residue may contain some carbonaceous materials, which changes its color from light gray to blackish. Therefore, in addition to metal oxide residue, the presence of LDH facilitates the carbonaceous char formation.

As far as the residue yield under nitrogen and oxygen atmosphere during TGA experiments is considered, comparable results are obtained in the case of filled systems (Fig. 23b). For the nanocomposite composition containing 8.93 wt % LDH, the comparison between the TGA analyses under nitrogen and under oxygen atmosphere is shown in Fig. 24. The nanocomposite undergoes a two-stepped decomposition in the presence of oxygen whereas under nitrogen atmosphere a single-stepped decomposition occurs with the low temperature decomposition maximum in the former case at around 430 °C being completely suppressed. It is obvious that the high amount of residue obtained under non-oxidative atmosphere indicates the presence of carbonaceous char. But, how can a similar amount of residue also be obtained under

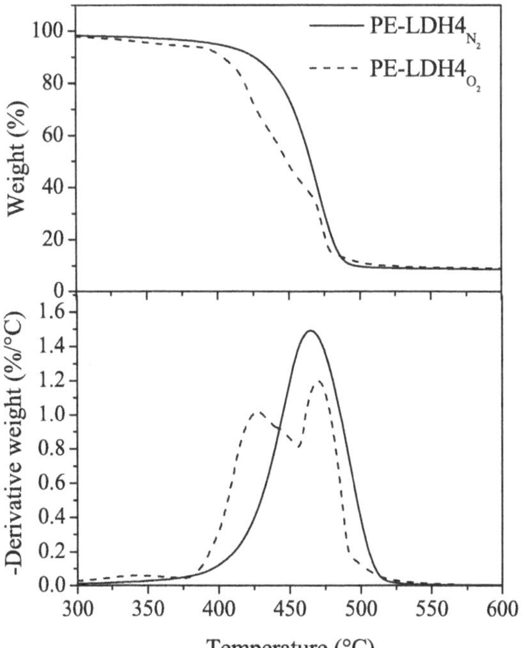

Fig. 24 Comparison of the TGA plots of LDPE/Mg – Al LDH nanocomposite under oxygen and nitrogen atmosphere (Reprinted from [135], with permission from Elsevier)

oxygen atmosphere? A possible explanation is the delayed or restricted access of oxygen in the interlayer region of LDH. This results in non-oxidative thermal decomposition of the organic contents present within the LDH layers. Additionally, the endothermic decomposition of the surrounding metal hydroxide layers causes a cooling effect, which may further affect the combustion process of these constraint organic species. As a result, a significant amount of carbonaceous char formation takes place even under oxygen atmosphere. In fact, such carbonaceous char formation within clay layers has been proposed during combustion of polymer/layered silicate-based nanocomposites [125]. The basic difference between layered silicate and LDH during combustion is that the layer undergoes endothermic decomposition liberating water and metal oxide, which may cause lower mechanical stability of the char than in the case of LDH.

4.5 Flammability Properties

4.5.1 Cone Calorimeter Investigation

The cone calorimeter test is an advanced and widely used test method for assessing flammability of polymeric materials. The most important parameter monitored during this test is heat release rate (HRR), which is calculated from the amount of oxygen consumed during combustion based on the principle described by Hugget [126]. During the whole cone calorimeter investigation, a constant external heat flux is maintained to sustain the combustion of the test sample i.e., the test method creates a forced flaming combustion scenario. Therefore, the test results from cone calorimeter are more significant for flammability evaluation of the specimen than the results obtained from tests that depend on a material's heat of combustion for sustaining the combustion process, like LOI and UL94 test methods. The LDPE/LDH nanocomposites were investigated following the standard ISO 5660 using a sample size of $100 \times 100 \times 4$ mm and an external heat flux of 30 kWm^{-2}; the results are summarized in Figs. 25 to 29.

HRR is a very important variable and represents how fast a fire can reach an uncontrollable stage. This single parameter provides information regarding the size of the fire and how fast it grows. The effectiveness of a fire retardant additive in a polymer can also be assessed with respect to this parameter. Figure 25 shows that increasing the concentration of LDH in

Fig. 25 Cone calorimeter investigation results showing **a** variation of heat release rate (*HRR*) with time, **b** variation of time of ignition (t_{ig}) and peak heat release rate (*PHRR*) with LDH content, and **c** total heat released (*THR*) with time in LDPE/Mg – Al LDH nanocomposites (Reprinted from [135], with permission from Elsevier) ▶

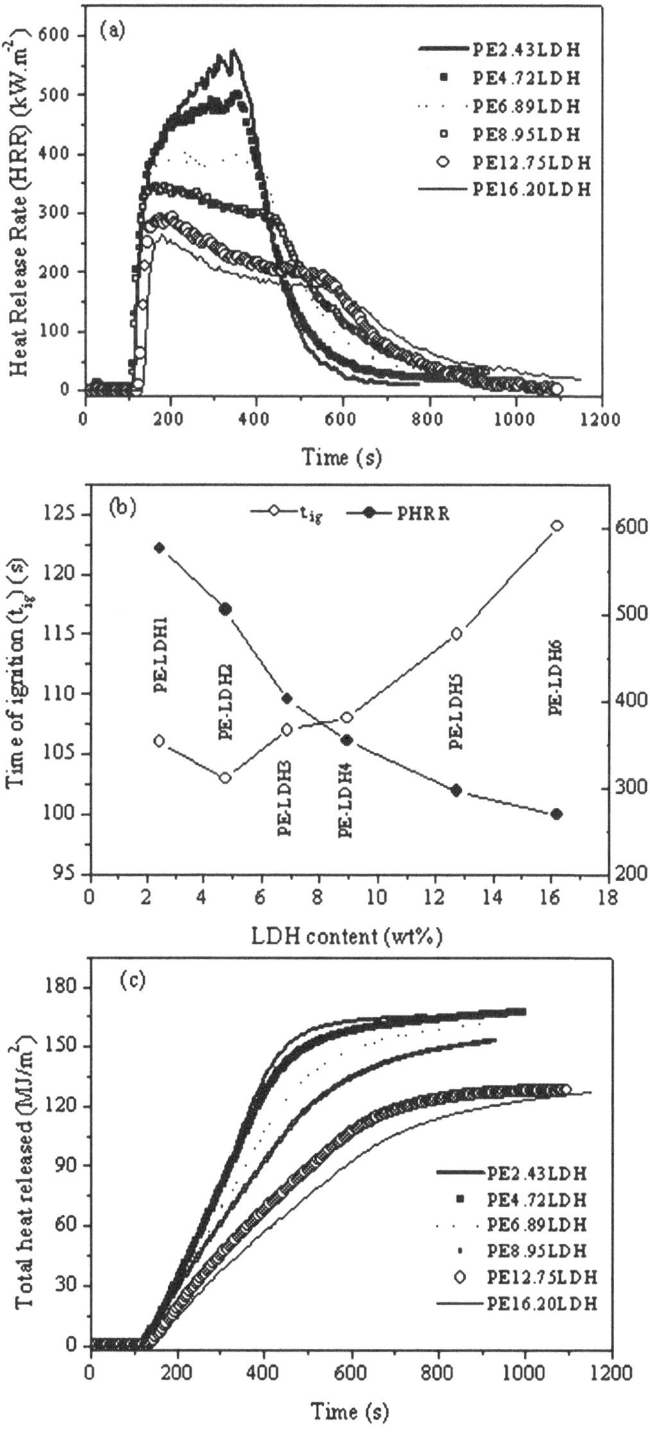
(a)
PE2.43LDH
PE4.72LDH
PE6.89LDH
PE8.95LDH
PE12.75LDH
PE16.20LDH
Heat Release Rate (HRR) (kW.m^{-2})
Time (s)
(b)
t_{ig}
PHRR
PE-LDH1
PE-LDH2
PE-LDH3
PE-LDH4
PE-LDH5
PE-LDH6
Time of ignition (t_{ig}) (s)
PHRR (kW.m^{-2})
LDH content (wt%)
(c)
PE2.43LDH
PE4.72LDH
PE6.89LDH
PE8.95LDH
PE12.75LDH
PE16.20LDH
Total heat released (MJ/m^2)
Time (s)

the nanocomposite not only significantly reduces the peak heat release rate (PHRR), but also makes the HRR curve increasingly flattened. This is caused by the decreasing burning rate of the materials with increasing LDH concentration. For unfilled polyethylene, cone calorimeter investigation under similar external heat flux (30 kWm^{-2} and 35 kWm^{-2}) gives a PHRR value over 800 kWm^{-2} [127, 128]. The addition of a small amount of LDH causes significant reduction in PHRR in LDPE (below 600 kWm^{-2}). At higher LDH concentration the PHRR is lowered below 300 kWm^{-2}. Ignition of the test specimen is followed by the formation of a molten surface layer on which the flame floats. As the combustion of the surface layer of the unfilled specimen leaves no residue, the molten layer is always directly exposed to the flame, causing fast decomposition of the material. But, in the nanocomposites, especially at higher LDH concentration, a compact layer of residue is formed that separates the flame region from the molten layer and acts as a barrier against the heat conduction from the former region to the latter. This results in much slower material decomposition in the nanocomposites.

The ignition time (t_{ig}) is defined as the time at which flaming is sustained over the entire surface of the specimen, and is also significantly increased with increasing LDH content. In the case of unfilled LDPE, t_{ig} is observed below 100 s, and increases to above 120 s with 16.20 wt % LDH content in the nanocomposites (Fig. 25b). This indicates that the resistance against catching fire is improved in the presence of LDH and with increasing LDH concentration. The total heat released (THR) is a parameter that indicates how big the fire is. Once the ignition takes place, THR steadily increases with burning time and attains a steady state before the flameout occurs. The THR, over the first 5 min and 10 min after application of an external heat flux, is progressively reduced with increasing LDH content (see Fig. 25c). At 10 min after the application of external heat flux, the THR value is reduced by about 17% and 44% in the samples with 6.89 and 12.72 wt % LDH, respectively, in comparison to the THR observed for the nanocomposite composition with lowest LDH content (2.43 wt %).

THR is often taken as the measure of the propensity to sustain a long duration fire. An efficient flame retardant should reduce THR considerably when incorporated into a polymer. In this respect, LDH fillers have definite advantage over the layered silicates as flame retardants. The presence of layered silicates in a polymer nanocomposite does not cause any significant change in the THR value in comparison to unfilled polymer, even if the concentration of the filler is increased [129]. This is because the layered silicates remain chemically inert during the burning process and hence do not influence the heat of combustion, which is roughly similar for unfilled polymer and the nanocomposites. They simply act as a physical barrier between the flame front and the burning surface. In contrast, LDH takes part actively in the combustion process through endothermic decomposition, acting as a heat sink and reducing the THR value during combustion. To measure how fast fire can grow during the burning process, a parameter equal to the ratio PHRR/t_{ig} is defined, which is related to the propensity to cause fast growth of fire. Like having low THR, an efficient flame retardant should also result in a low value of this parameter to obtain a flame retardant composite. In fact, when THR is plotted against PHRR/t_{ig}, the position of a good flame retardant composite would be close to the origin [129]. Figure 26 shows comprehensive assessment of LDPE/LDH nanocomposites using this principle. It can be that increasing LDH concentration not only reduces the THR, but also the PHRR/t_{ig} ratio. Beyond a certain LDH concentration, the position of the nanocomposites in THR versus the PHRR/t_{ig} co-ordinate system sharply bends towards the origin with increasing LDH concentration. This is certainly a promising development when compared to the performance of layered silicate-based nanocomposites. In the case of polypropylene/layered silicate-based composites, it has been observed that with increasing filler concentration the PHRR/t_{ig} ratio decreases significantly. But, since the THR value remain roughly constant, the positions of this system lie parallel to the PHRR/t_{ig} axis in similar plots to those in Fig. 26.

Figure 27 exhibits the influence of LDH loading on rate of carbon dioxide (CO_2) and carbon monoxide (CO) emission during combustion. During the first 500 s of combustion in the cone calorimeter, the release rates of CO_2 and CO are reduced significantly in the nanocomposite composites with increasing LDH concentration. But, beyond 500 s, the unfilled LDPE and the LDPE/LDH compositions with low LDH concentration show a drastic drop in

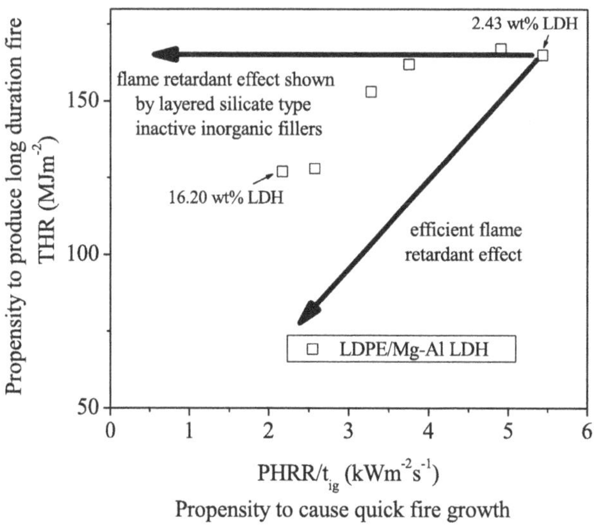

Fig. 26 Graphical assessment of fire risk associated with PE/LDH nanocomposites, plotting THR against PHRR/t_{ig} for different LDH concentration. The LDH concentration increases from *right to left* (Reprinted from [135], with permission from Elsevier)

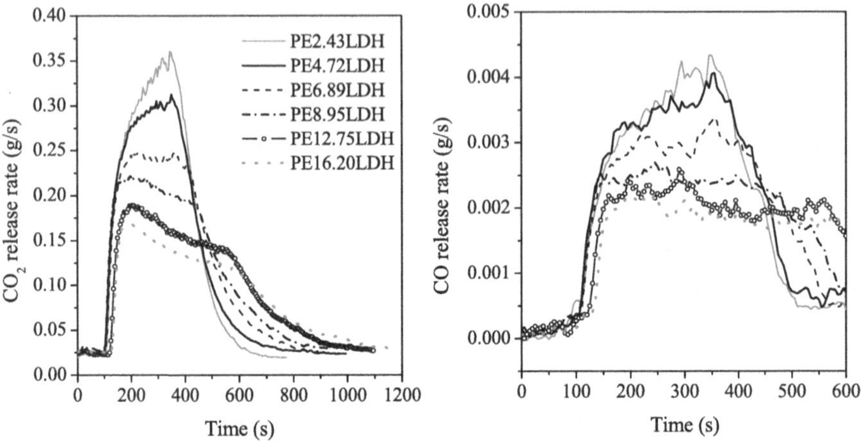

Fig. 27 Variation of carbon dioxide (CO_2) and carbon monoxide (CO) release rate during combustion in a cone calorimeter chamber with burning time

the release rates of these two gases. This is attributed to the increase in the burn time with increasing LDH concentration, which results in a steady but prolonged release of these two gases from the nanocomposites. For example, the total burn time (the time at which flame-out takes place during cone calorimeter combustion test) is found to be about 660 s, and above 1000 s for the nanocomposites containing 2.43 wt % and 16.20 wt % LDH, respectively. Another possible explanation of the increasing CO release rate during the later stage of burning in Fig. 27 is the slow thermo-oxidative reaction of the carbonaceous residue. In fact, a similar effect has also been observed in the case of carbon nanotube filled polymer nanocomposites [130].

The yields of residue can be estimated directly from the cone calorimeter tests. Based on these results and the assumption that polyethylene burns

Table 4 Comparison of total residue and carbonaceous residue formed after combustion of LDPE/Mg – Al LDH nanocomposites in the cone calorimeter chamber

Hydroxide content LDPE/ Mg – Al LDH compositions wt%	Combustion residue Calculated wt%	Cone calorimeter wt%	Carbonaceous residue wt%
2.43	1.79	2.77	0.98
4.72	3.47	5.10	1.63
6.89	5.07	11.80	6.73
8.95	6.59	14.25	7.66
12.75	9.39	19.97	10.58
16.20	11.92	27.97	16.05
LDPE + 15 wt % MH	10.35	18.01	7.67

completely leaving no residue, the amount of carbonaceous char can be calculated using the similar decomposition reaction used for LDH-DBS earlier. The results are summarized in Table 4. All the nanocomposite samples produce a higher amount of char than the theoretical values, which are based only on the metal oxide content of the residue. The difference increases with increasing LDH concentration. This means that the yield of carbonaceous char increases with increasing LDH concentration. The comparison between LDH-filled and MH-filled systems with comparable filler loading shows that the latter yields a much lower amount of char. This indicates that the LDH is a more efficient char former during combustion than the simple MH.

4.5.2
LOI and UL94 Investigation

Evaluation of the composite materials in terms of limited oxygen index (LOI) provides first hand information on the effectiveness of fire retardant additives [22]. For LDPE/LDH nanocomposites with up to 16.20 wt % LDH content, the LOI values are shown in Fig. 28. The unfilled LDPE has a LOI value of 18.0 and the addition of a small amount of LDH has nearly no effect on the LOI value. Even a small decrease in the LOI value is observed below 5 wt % LDH concentration. However, at higher levels of LDH content (above 5 wt %), the LOI value increases with increasing LDH concentration. The nanocomposites containing 16.20 wt % LDH have a LOI value around 22.0, which is significantly higher than that of unfilled LDPE, but still further improvement is necessary for industrial acceptance.

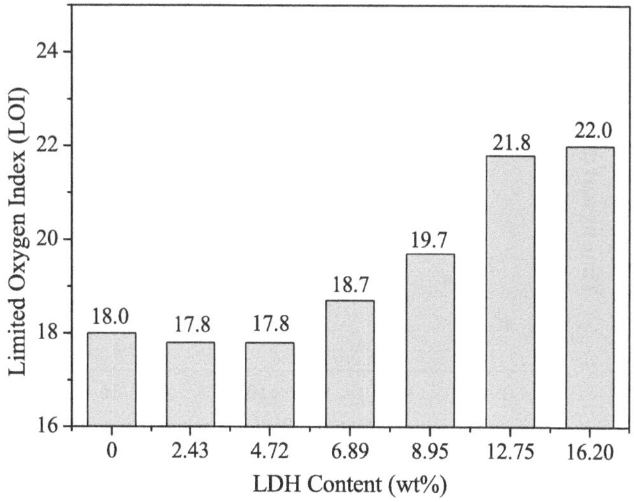

Fig. 28 Limited oxygen index of PE/LDH nanocomposites (Reprinted from [135], with permission from Elsevier)

UL94 testing was carried out using two different standards: the vertical burn test (UL94V) and the horizontal burn test (UL94HB). The LDPE/LDH nanocomposites containing up to 16.20 wt % LDH did not pass any of the UL94V specifications. All the samples start burning spontaneously after the first 10 s flame application, which continued until the test specimen was completely burnt up to the sample-holding clamp. Nevertheless, the UL94V burn test provides useful information regarding the dripping behavior of these composite materials. Dripping of the burning melts directly influences the spread of flame through secondary flaming during real-life burning situations. Although all the samples drip while burning, the time at which dripping starts is significantly delayed by the presence of LDH. A parameter called "time to start dripping" was defined as the time after the first 10 s flame application at which the first dripping occurs that ignites a piece of cotton kept underneath. As shown in Fig. 29, the time to first dripping increases steadily with increasing LDH concentration in the nanocomposites and becomes more than ten times in the case of PE-LDH6. A similar trend was also observed in the case of polypropylene/layered silicate-based nanocomposites, where none of the composition achieved any UL94V test rating, but showed improved dripping behavior compared to unfilled polypropylene [131]. The high melt viscosity of the nanocomposite, which increases with increasing LDH concentration, causes the slow and much-delayed dripping in LDPE/LDH nanocomposites. While in the case of unfilled LDPE dripping starts in the form of a continuous stream of burning melt, the dripping in the nanocomposite takes place in the form of chunks of glowing burn residue, especially at high LDH concentrations.

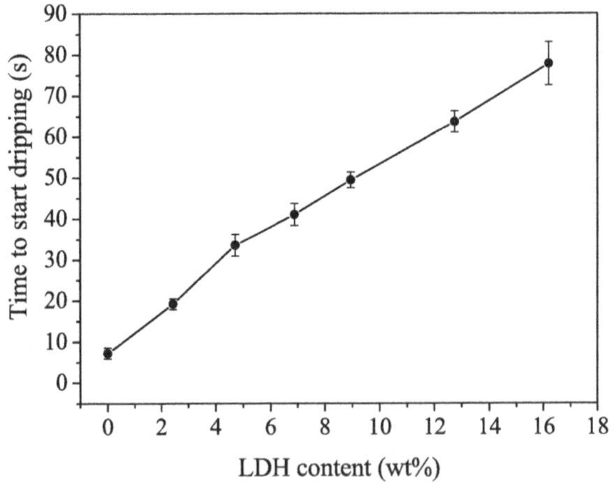

Fig. 29 Effect of LDH loading on the dripping tendency of the polyethylene/LDH nanocomposites (time to start dripping is the time after the first 10 s flame application at which the first dripping occurs that ignites a piece of cotton underneath during UL94V testing)

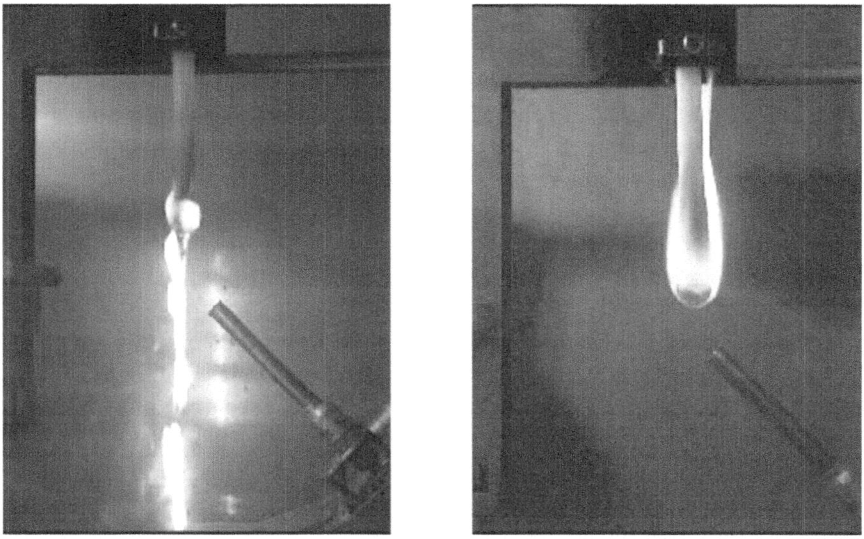

Fig. 30 Dripping behavior of unfilled LDPE (*left*) and PE/LDH nanocomposite containing 12.75 wt % LDH (*right*). Pictures were taken after approximately the same time period from the cessation of first 10 s flame application during UL94V test

Figure 30 shows typical dripping behavior of unfilled LDPE and a LDPE/LDH composition (PE-LDH5) at comparable time after the first 10 s flame application. The UL94 horizontal burn test showed better flammability performance of the LDPE/LDH nanocomposites as compared to the pure polyethylene.

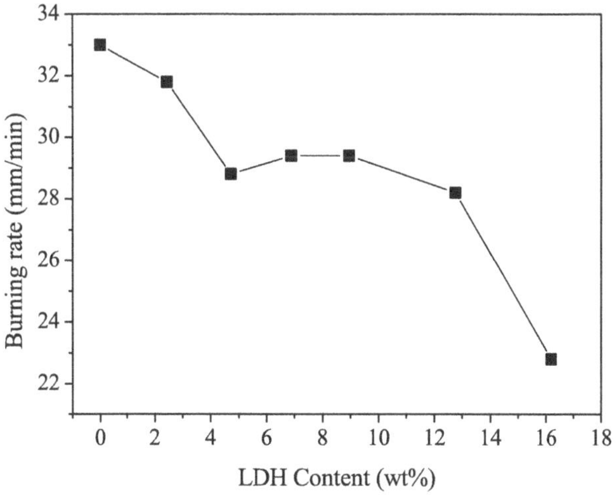

Fig. 31 Influence of LDH loading on the rate of burning during UL94 horizontal burn test (at each composition, the average of three measurements were taken) (Reprinted from [135], with permission from Elsevier)

The nanocomposite compositions containing 4.72 wt % or higher LDH easily passed the UL94HB rating. This test method is also useful to determine the rate of burning (expressed as mm min^{-1}) of the testing material. In LDPE/LDH nanocomposites, increasing the LDH concentration significantly delays the burning rate, as can be seen in Fig. 31. When LDH concentration is increased above 10 wt %, the specimen burns extremely slowly.

4.5.3 LDH as Flame Retardant Synergist

Although LDPE/LDH nanocomposites show significant improvement in flammability properties, they cannot provide industrially acceptable LOI values and UL94 test ratings, especially at low LDH concentrations. LDHs are basically metal hydroxides, which exhibit their flame retardant effect mainly through the endothermic decomposition and barrier formation by the metal oxide residue. Because of their mechanism of flame inhibition, they are far less efficient as flame retardants than the halogenated flame retardants, which directly deactivate the flame-propagating reactive radical species in the flame zone. However, LDH can play a role of synergist in a flame retardant package, where its char-forming ability and active nature can be helpful in reducing the total amount of flame retardant required for satisfactory flammability ratings. For example, Zammarano et al. [132] observed a synergistic effect between Mg – Al LDH and ammonium polyphosphate (APP) in epoxy-based composites. It was found that with the addition of 4 wt % organically modified LDH, the concentration of APP can be reduced from 30 wt % to 16–20 wt % to obtain the UL94V0 rating. Similar improvement in flame retardancy has also been observed between conventional metal hydroxide-type flame retardants and layered silicate-type nanofillers. For example, in an ethylene vinyl acetate (EVA)/aluminum trihydrate (ATH)-based compound the concentration of ATH can be reduced from 65 wt % to 45 wt % with incorporation of 5 wt % organically modified layered silicate [133]. Such replacement reduces the total filler concentration significantly (thus, maintaining or improving the mechanical properties and processibility of the final composites) and at the same time maintains the heat release rate at about 200 kW m^{-2}. In the case of LDPE/MH-based composites, similar synergistic effect has also been observed when MH is partially replaced by organically modified LDH. Typically, in polyolefin composites containing MH only, a filler concentration higher than 60 wt % is often required to obtain satisfactory flammability rating. Such a high concentration of filler not only severely deteriorates the mechanical properties of the composites, but also affects the processibility of the melt. Therefore, the aim of introducing LDH in parts in these composites is to improve the dispersion of the filler and also give reinforcement of the melt, which improves the dripping resistance during burning. The synergistic effect between organically modified LDH (LDH-DBS) and MH (MH) was investigated by comparing the thermal and flammability properties of the LDPE/MH control composition with those of the LDPE/LDH/MH compositions.

The LOI vales of LDPE/MH and LDPE/LDH/MH composites are compared in Fig. 32. The three control LDPE/MH compositions are based on 40, 50 and 60 wt % of MH. The maximum amount of LDH-DBS has been kept at 10 wt % in the combined systems. As the proportion of LDH in the combined system is increased, the LOI value increases compared to the corresponding control composition. The compositions LDPE/10LDH/30MH and LDPE/10LDH/40MH show LOI values comparable to those found for the control compositions LDPE/50MH and LDPE/60MH, respectively. This means that overall filler loading in a LDPE/MH composite can be significantly reduced in the presence of LDH to obtain a desired LOI value. The explanation of such synergism can be obtained if the burning process is followed carefully. In the case of the composites containing only MH at concentrations below 60 wt %, the burn residue/char is not held firmly with the sample stock. Therefore, as the char grows in size, it falls easily under its own weight exposing a fresh surface for burning. The presence of LDH enhances the melt viscosity and also the strength of the char. This provides sufficient mechanical stability to the char so that it is held firmly at the tip of the LOI test specimen. The char layer cuts off the direct contact between the flame region and the unburnt layer under the char. As a result, flame is extinguished and a higher concentration of oxygen is required to sustain the combustion. Also, the improved dispersion of LDH particles causes better distribution of cooling effects during burning throughout the matrix.

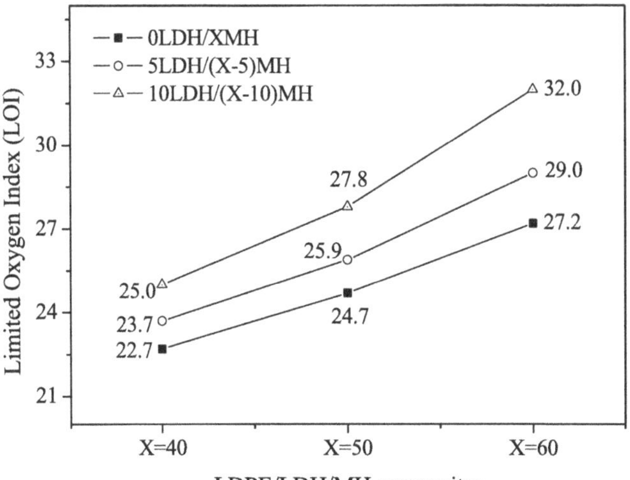

Fig. 32 Synergistic effect of LDH in MH-filled polyethylene composites. The *numbers* near data points indicate the exact LOI value and those in the legend indicate the amount of LDH or MH in wt%; *X* represents total wt% of filler, i.e., MH + LDH

Thermogravimetric analysis shows that the thermo-oxidative degradation of PE/MH composites containing 40 wt % MH starts at about 375 °C and continues up to about 475 °C (Fig. 33). The decomposition takes place in two distinct stages characterized by two decomposition maxima in the differential plot at about 400 °C and 460 °C (Fig. 33). During the first decomposition stage, which spans over the range 375–415 °C, LDPE/MH composite (containing 40 wt % MH) loses less than 25% of its weight. By about 415 °C, the unfilled LDPE and MH individually lose about 70% and 28% of their respective weights. So, principally, LDPE/MH composite containing 40 wt % MH would lose about 53 wt % ($\approx 0.70 * 60 + 0.28 * 40$) of its weight by this temperature. But, in practice, Fig. 33 shows that the loss is about 25 wt %. Therefore, it can be concluded that the endothermic decomposition of MH largely suppresses

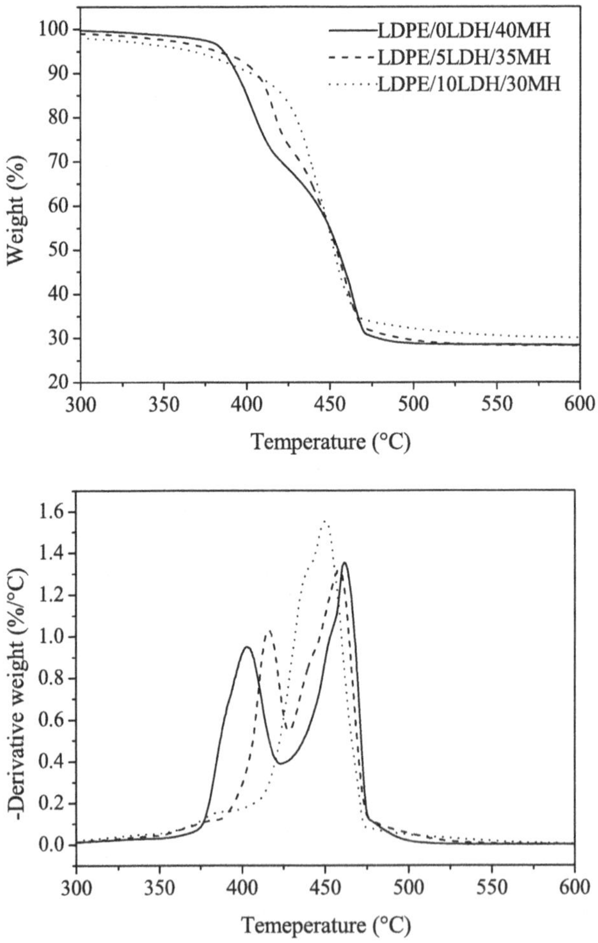

Fig. 33 Thermogravimetric analysis of PE/MH and PE/LDH/MH composites

the first decomposition peak of the polyethylene matrix in LDPE/MH and that most of the matrix decomposition takes place above 415 °C. When MH in LDPE/MH composite is partially replaced by increasing amounts of LDH, significant changes in the thermal decomposition behavior are observed. The first decomposition peak, which corresponds to the decomposition of filler and in part of the matrix, shifts to higher temperatures with increasing LDH proportions and ultimately merges with the second decomposition stage. This means that the two-stage decomposition observed in LDPE/40MH is changed to a single-stage decomposition in LDPE/10LDH/30MH. This is a clear indication of improvement of thermal stability of the composites in the presence of LDH.

The flammability performance of LDPE/MH and LDPE/LDH/MH composites has also been compared using UL94 vertical burn test results, which are summarized in Table 5. The test samples with specified dimensions (125 mm × 10 mm × 4 mm) were tested for both UL94 vertical and horizontal ratings. In UL94V testing, the test specimen is exposed to flame twice, each time for 10 seconds. It is apparent that partial replacement of MH by LDH also improves the performance of the materials in the UL94 vertical burn test. The composites containing 40 wt % total filler content (MH + LDH) even after introduction of 10 wt % LDH do not show positive UL94 ratings. However, the dripping resistance of the material improves significantly as more and

Table 5 Summary of UL94 vertical burn test for PE/MH and PE/LDH/MH composites (sample size: 125 mm × 10 mm × 4 mm)

Sample	LDH wt%	MH wt%	t_1[a] s	Dripping	t_2[a] s	Dripping	t_3[b] s	UL94 rating
LDPE/0LDH/40MH	0	40	A, B	Yes	–	–	68[c]	No
LDPE/5LDH/35MH	5	35	A, B	Yes	–	–	125[c]	No
LDPE/10LDH/30MH	10	30	8	No	A, B	Yes	83[d]	No
LDPE/0LDH/50MH	0	50	A, B	Yes	–	–	151[c]	No
LDPE/5LDH/45MH	5	45	11	No	A, B	Yes	66[d]	No
LDPE/10LDH/40MH	10	40	2	No	4	No	–	V0
LDPE/0LDH/60MH	0	60	2	No	6	No	–	V0
LDPE/5LDH/55MH	5	55	0	No	2	No	–	V0
LDPE/10LDH/50MH	10	50	0	No	0	No	–	V0

[a] t_1 and t_2 Burning time after first and second heating, respectively
[b] t_3 Time at which dripping starts
[c] t_3 Counted from the end of first heating
[d] t_3 Counted from the end of second heating
A drips, and the cotton below burns; *B* sample burns up to the sample holder

more LDH is incorporated. The sample containing 40 wt % MH shows continuous melt dripping after first heating. While the sample containing 30 wt % MH and 10 wt % LDH (10L30MH) does not show any melt dripping after first heating, but the flame is extinguished after some time. However, this sample burns continuously and shows dripping after the second heating step. An MH content up to even 50 wt % is not sufficient for a positive UL94V rating and the sample burns continuously with melt dripping after first heating. When 20% of MH in PE/50MH is replaced by LDH (i.e., 10L40MH), a positive UL94V rating with V0 classification can be achieved. All the compositions containing 60 wt % total filler are self-extinguishing and show UL94V0 rating. When MH is replaced by LDH, the burning time after each heating step in UL94V testing is decreased.

The main purpose of substituting MH with Mg – Al LDH in polyethylene-based composites was to reduce the overall metal hydroxide content to obtain satisfactory flame retardancy. But, in doing so, one should not underestimate the effects on other properties, such as mechanical properties and processibility. In fact, a big compromise in these properties would not be acceptable for useful applications. Therefore, an increase in proportion of LDH in a MH/LDH combination is limited by the processibility and the mechanical properties of the composites. Table 6 shows a summary of the mechanical properties of the various LDPE/LDH/MH compositions investigated in the present work. The incorporation of LDH up to 10 wt % in presence of maximum 50 wt % MH does not deteriorate the modulus and yield strength of the composites, rather, a small increase in both the properties can be observed. However, at 10 wt % LDH content in PE/LDH/MH composites containing high amounts of MH (above 40 wt %), the elongation properties of final composites are affected significantly indicating a lowering of the impact strength of

Table 6 Summary of mechanical properties of PE/MH and PE/LDH/MH composites

Sample	Modulus MPa	SD –	Yield strength MPa	SD –	EB %	SD –
LDPE/0LDH/40MH	379.6	16.9	12.64	0.09	39.57	3.24
LDPE/5LDH/35MH	368.0	22.6	13.19	0.21	34.51	2.96
LDPE/10LDH/30MH	424.9	24.2	13.36	0.25	21.29	2.46
LDPE/0LDH/50MH	540.4	59.0	12.73	0.15	18.00	3.76
LDPE/5LDH/45MH	484.1	15.8	13.76	0.13	19.49	1.69
LDPE/10LDH/40MH	539.2	36.4	13.72	0.19	10.06	0.86
LDPE/0LDH/60MH	672.9	42.2	13.71	0.26	9.25	1.20
LDPE/5LDH/55MH	682.8	60.6	14.55	0.29	7.48	0.73
LDPE/10LDH/50MH	741.5	47.4	14.45	0.33	4.65	0.57

SD standard deviation; *EB* elongation at break

the materials. The use of some impact modifier (such as ethylene propylene diene (EPDM) rubber or ethylene vinyl acetate) might be effective in obtaining better mechanical properties at higher LDH substitution.

5 Conclusions

The potential of LDH, especially of Mg – Al LDH, as a new kind of nanofiller for the synthesis of polymer-based nanocomposites has been discussed in this article. Several advantages of LDH materials have been highlighted that show on the one hand the possibility of addressing the dispersion problems associated with conventional inorganic fillers, like metal hydroxides, and on the other hand project them as competitive nanofillers to commonly used clay nanofillers, like layered silicates. The easy synthetic procedure, cationic charge of the metal hydroxide layers, chemically active nature, etc. are the various aspects that raise serious interest about LDH materials. The synthetic procedures such as co-precipitation followed by a suitable hydrothermal treatment can not only produce highly crystalline materials of nanoscopic particle size, but also provide the possibility of synthesizing LDHs based on various combinations of metal ions. The cationic layer charge makes LDHs suitable for modification by a wide range of anionic surfactants, which are often cheap and industrially available. The chemically active nature is helpful in direct participation of these materials during flame inhibition, thus improving the flame retardancy of the composite materials. Because of these advantages, the synthesis of LDH-based polymer nanocomposites is a rapidly developing field in polymer composite research. An overview on the different synthetic methods and various LDH-based polymer nanocomposite systems reported in recent years has been provided in the present article.

Organic modification of LDHs is often necessary to convert them into suitable precursors for synthesis of polymer nanocomposites. Being an anionic clay, the chemicals for modification of LDHs can be selected from a wide range of organic and inorganic materials, like anionic surfactants, oxo- and polyoxometallates anions, oraganometallic complex anions, etc. In the present study, a number of anionic surfactants were used to modify Mg – Al LDH in order to enlarge the interlayer distance and to render it more organophilic. The surfactants were selected according to their functionality, chain length, etc. XRD analysis of the modified clay reveals that the surfactant anions are arranged as a monolayer within the interlayer region of LDH clay and increase the interlayer distance proportionally to the length of their hydrocarbon chain. Some water molecules are also accommodated in the interlayer region in the modified LDH, and are usually arranged as a monolayer between the surfactant's tail and the metal hydroxide sheet. However, in the case of the surfactant having a more bulky hydrocarbon tail (e.g., two hydrocarbon chains as in BEHP), an absence of interlayer water is more likely. From the various functionalities of the surfactants, LDH shows higher affinity to the sulfate and sulfonate type of anionic functionalities, which has been confirmed by their XRD pattern. For example, in the case of LDH-DS and LDH-DBS, no reflection maximum corresponding to the unmodified clay can be observed. However, with phosphate and carboxylate, the formation of some unmodified LDH during the regeneration process is indicated by the presence of its characteristic XRD reflection in the modified materials. The original plate-like particle morphology of the unmodified LDH materials is not changed after organic modification except for a higher surface roughness and disordered edges. The intercalation of bigger anionic species into the interlayer space results in swelling of the LDH platelets. Thermal analysis reveals that the unmodified LDH show large decomposition peaks below 300 °C due to complete loss of the interlayer water molecules and a partial loss of the carbonate anions. The organic modification, especially with DBS, either significantly suppresses these peaks or shifts them to a higher temperature. The modified LDH materials, in addition to modifying surfactant anions, also contain some interlayer water and carbonate anions that are incorporated during the regeneration process.

To investigate the potential of Mg – Al LDH as a flame retardant nanofiller, polyethylene-based nanocomposite has been investigated in detail. It is still a challenge to obtain a high degree of exfoliation of inorganic clay materials in a polyolefin matrix through melt compounding processes. Using Mg – Al LDH we observed partial exfoliation with the formation of localized networked structures both at the micro- and nanoscale. The primary particles, representing a single LDH platelet built from a large number of metal hydroxide sheets, not only undergo breakdown into smaller fragments through delamination of thinner stacks from their surface, but also form aggregates that look like domains of closely associated platelets. The exfoliated fragments or clay layers either remain scattered in the matrix or also form localized regions of network structure. Overall, a hierarchy of structures from the nanoscopic-exfoliated layers to microscopic aggregated structures is observed in the LDPE/LDH nanocomposite. This particle morphology, along with possible adsorption of polymer chain segments on the LDH particle surface, significantly influences the melt rheological behavior of the nanocomposite melts. The dynamic oscillatory shearing in the linear viscoelastic regime reveals that with increasing LDH concentration, the melts progressively deviate from a low-frequency Newtonian behavior to a strong shear-thinning behavior. Also, the pseudo-solid-like behavior becomes more pronounced at high concentrations of LDH with the appearance of an apparent yield stress. The effect of localized network structure formed by the LDH particles is also reflected during a non-linear shear response. The increasing LDH concentration not only results in the appearance of a strong stress overshoot peak (due to the elastic response of the filler structures) at the flow inception during steady shear, but also shifts the stress overshoot peak assigned to the matrix phase to a lower shear deformation than that observed in unfilled melt. This behavior has been quantitatively addressed in the present study.

This report shows that LDHs, especially Mg – Al LDH, have a definite potential as flame retardants for a polymer matrix. With a highly flammable matrix, like polyethylene, significant improvements in flame retardancy have been observed. Although LDHs alone may not be as effective as halogen-containing flame retardants in obtaining a satisfactory flammability rating according to various industrial standards, their combination with conventional flame retardants like metal hydroxide, ammonium polyphosphate, etc. is found to provide a synergistic effect, thereby noticeably reducing the overall flame retardant concentration in the final composites. Such combinations can be very innovative in developing a halogen-free, non-toxic flame retardant system for many polymer composites. Besides, the nanoscopic dispersion of one of the ingredients of these flame retardant combinations can also provide mechanical reinforcement of the matrix. Additionally, there exists a tremendous potential of LDH-based nanohybrid materials in various other fields of application. The choice of suitable anionic species for the modification of LDHs plays a key role in designing hybrid materials with specialized properties.

References

1. LeBaron PC, Wang Z, Pinnavaia TJ (1999) Appl Clay Sci 15:11
2. Alexandre M, Dubois P (2000) Mater Sci Engg 28:1
3. Sinha Ray S, Okamoto M (2003) Prog Polym Sci 28:1539
4. Leuteritz A, Kretzschmar B, Pospiech D, Costa FR, Wagenknecht U, Heinrich G (2007) Industry-relevant preparation, characterization, and applications of polymer nanocomposites. In: Nalwa HS (ed) Polymeric nanostructures and their applications. American Scientific Publishers, Los Angeles
5. Tichit D, Coq B (2003) Cattech 7:206
6. Cavani F, Trifiro F, Vaccari A (1991) Catal Today 11:173
7. Choy JH, Kwak SY, Jeong YJ, Park JS (2000) Angew Chem 111:4207
8. Hoyo CD (2007) Appl Clay Sci 36:103
9. Van Der Ven L, Van Gemert MLM, Batenburg LF, Keern JJ, Gielgens LH, Koster TPM, Fischer HR (2000) Appl Clay Sci 17:25
10. Miyata S, Hirose T, Iizima N (1978) US Patent 4085088
11. Du L, Qu B, Zhang M (2007) Polym Degrad Stab 92:497
12. Cardoso LP, Celis R, Cornejo J, Valim JB (2006) J Agric Food Chem 54:596
13. Lakraimi M, Legrouri A, Barroug A, De Roy A, Besse JP (1999) J Chim Phys 96:470
14. Zhuravleva NG, Eliseev AA, Lukashin AV, Kynast U, Tret'yakov UD (2004) Doklady Chem 396:60
15. Chang Z, Evans D, Duan X, Boutinaud P, de Roy M, Forano C (2006) J Phys Chem Solids 67:1054
16. Guo S, Evans DG, Li D (2006) J Phys Chem Solids 67:1002
17. Tian Y, Wang G, Li F, Evans DG (2007) Mater Lett 61:1662

18. Lukashin AV, Vertegel AA, Eliseev AA, Nikiforov MP, Gornert P, Tretyakov UD (2003) J Nanopart Res 5:455
19. Mohana D, Pittman CU Jr (2007) J Hazard Mater 142:1
20. Lv L, He J, Wei M, Evans DG, Zhou Z (2007) Water Res 41:1534
21. Leroux F, Besse JP (2001) Chem Mater 13:3507
22. Lomakin SM, Zaikov GE (2003) Modern polymer flame retardancy. VSP, Netherlands
23. Gilman JW, Jackson CL, Morgan AB, Harris R Jr (2000) Chem Mater 12:1866
24. Cavani F, Trifiro F, Vaccari A (1991) Catalysis Today 11:173
25. Reichle WT (1986) Solid State Ionics 2:135
26. Meyn M, Beneke K, Legaly G (1990) Inorg Chem 29:5201
27. Clause O, Gazzano M, Trifiro F, Vaccari A, Zatroski L (1991) Appl Catal 73:217
28. Sychev M, Prihod'ko R, Erdmann K, Mangel A, van Santen RA (2001) Appl Clay Sci 18:103
29. König U (2006) PhD Thesis, Martin-Luther-Universität Halle-Wittenberg, Germany
30. Costantino U, Marmottini F, Nocchetti M, Vivani R (1998) Eur J Inorg Chem, p 1434
31. Ogawa M, Kaiho H (2002) Langmuir 18:4240
32. Mascolo G (1995) Appl Clay Sci 10:21
33. Khan AI, O'Hare D (2002) J Mater Chem 12:3191
34. Malki KE, Guenane M, Forano C, De Roy A, Besse JP (1992) Mater Sci For 171:91–93
35. Bontchev RP, Liu S, Krumhansi JL, Voigt J, Nenoff TM (2003) Chem Mater 15:3669
36. Carlino S (1997) Solid States Ionics 98:73
37. You Y, Zhao H, Vance GF (2002) J Mater Chem 12:907
38. Prevot V, Forano C, Besse JP (2001) Appl Clay Sci 18:3
39. Miyata S (1980) Clays Clay Minerals 24:50
40. Meyn M, Beneke K, Legaly G (1990) Inorg Chem 29:5201
41. Buniak GA, Schreiner WH, Mattoso N, Wypych F (2002) Langmuir 18:5967
42. Bish DL, Brindley GW (1977) Amer Min 62:458
43. Costa FR, Leuteritz A, Wagenknecht U, Jehnichen D, Häußler L, Heinrich G (2007) Appl Clay Sci (in press) doi:10.1016/j.clay.2007.03.006
44. Borja M, Dutta PK (1992) J Phys Chem 96:5434
45. Messersmith PB, Stupp SI (1995) Chem Mater 7:454
46. Oriakhi CO, Farr IV, Lerner MM (1996) J Mater Chem 6:103
47. Wilson OC Jr, Olorunyolemi T, Jaworski A, Borum L, Young D, Siriwat A, Dickens E, Oriakhi C, Lerner M (1999) Appl Clay Sci 15:265
48. Whilton NT, Vickers PJ, Mann S (1997) J Mater Chem 7:1623
49. Leroux F, Aranda P, Besse JP, Ruiz-Hitzky E (2003) J Inorg Chem 6:1242
50. Vaysse C, Guerlou-Demourgues L, Duguet E, Delmas C (2003) Inorg Chem 42:4559
51. Lee WF, Chen YC (2004) J Appl Polym Sci 94:2417
52. Moujahid EM, Besse JP, Leroux F (2002) J Mater Chem 12:3324
53. Besse JP, Moujahid EM, Dubois M, Leroux F (2002) Chem Mater 14:3799
54. Tanaka M, Park IY, Kuroda K, Kato C (1989) Bull Chem Soc Jpn 62:3442
55. Isupov VP, Chupakhina LE, Ozerova MA, Kostrovsky VG, Poluboyarov VA (2001) Solid States Ionics 231:141–142
56. Vieille L, Taviot-Gueho C, Besse JP, Leroux F (2003) Chem Mater 15:4369
57. Vieille L, Moujahid EM, Tavoit-Gueho C, Cellier J, Besse JP, Leroux F (2004) J Phys Chem Solid 65:385
58. Roland-Swanson C, Besse JP, Leroux F (2004) Chem Mater 16:5512
59. Wang GA, Wang CC, Chen CY (2005) Polymer 46:5065
60. Wang GA, Wang CC, Chen CY (2005) J Inorg Organometal Polym Mat 15:239
61. Challier T, Slade RCT (1994) J Mat Chem 4:367

62. Sugahara Y, Yokoyama N, Kuroda K, Kato C (1988) Ceram Inter 14:163
63. O'Leary S, O'Hare D, Seely G (2002) Chem Commun 14:1506
64. Li B, Hu Y, Chen Z, Fan W (2007) Mater Lett 61:2761
65. Hsueh HB, Chen CY (2003) Polymer 44:1151
66. Chen W, Qu B (2003) Chem Mater 15:3208
67. Chen W, Feng L, Qu B (2004) Chem Mater 16:368
68. Qui L, Chen W, Qu B (2005) Polym Degrad Stab 87:433
69. Liao CS, Ye WB (2003) J Polym Res 10:241
70. Hsueh HB, Chen CY (2003) Polymer 44:5275
71. Li B, Hu Y, Liu J, Chen Z, Fan W (2003) Coll Polym Sci 281:998
72. Yang QZ, Sun DJ, Zhang CG, Wang XJ, Zhao WA (2003) Langmuir 19:5570
73. Costa AS, Imae T, Takagi K, Kikuta K (2004) Prog Collod Polym Sci 128:113
74. Leroux F, Aranda P, Besse JP, Hitzky ER (2003) Eur J Inorg Chem 6:1242
75. Miyata S (1983) US Patent 4 379 882
76. Nichols KL, Chou CJ (1999) US Patent 5 952 093
77. Costa FR, Abdel-Goad M, Wagenknecht U, Heinrich G (2005) Polymer 46:4447
78. Zammarano M, Bellayer S, Gilman JW, Franceschi M, Beyer FL, Harris RH, Meriani S (2006) Polymer 47:652
79. Lee WD, Im SS, Lim HM, Kim KJ (2006) Polymer 47:1364
80. Ding P, Qu B (2006) Polym Engg Sci 10:1153–1159
81. Du L, Qu B (2006) J Mater Chem 16:1549
82. Costantino U, Gallipoli A, Nocchetti M, Camino G, Bellucci F, Frache A (2005) Polym Degrad Stab 90:586
83. Morgan AB, Gilman JW (2003) J Appl Polym Sci 87:1329
84. Vaia RA, Liu W (2002) J Polym Sci Part B: Polym Phys 40:1590
85. Wang K, Liang S, Du R, Zhang Q, Fu Q (2004) Polymer 45:7953
86. Dennis HR, Hunter DL, Chang D, Kim S, White JL, Cho JW, Paul DR (2001) Polymer 42:9513
87. Lertwimolnun W, Vergnes B (2005) Polymer 46:3462
88. Solin SA, Hines D, Yun SK, Pinnavaia TJ, Thorpe MF (1995) J Non-Cryst Solids 182:212
89. Barnes HA (1997) J Non New Fluid Mech 70:1
90. Leonov AI (1990) J Rheol 34:1039
91. Macosko CW (1993) Rheology principles, measurements and applications. VCH, USA
92. Mezger TG (2002) In: Curl R (ed) The rheology handbook for users of rotational and oscillatory rheometer. Vincentz, Hannover
93. Ren J, Silva AS, Krishnamoorti R (2000) Macromolecules 33:3739
94. Baumgärtel M, Winter HH (1989) Rheol Acta 28:511
95. Winter HH, Mours M (2006) Rheol Acta 45:331
96. Liu X, Qian L, Shu T, Tong Z (2003) Polymer 44:407
97. Sobhanie M, Isayev AI (1999) J Non-Newt Fluid Mech 85:189
98. Larson RG (1999) The structure and rheology of complex fluids. Oxford University Press, New York
99. Costa FR, Satapathy BK, Wagenknecht U, Weidisch R, Heinrich G (2006) Eur Polym J 42:2140
100. Costa FR (2007) PhD thesis. Technische Universität Dresden, Germany
101. Witten TA, Leibler L, Pincus PA (1990) Macromolecules 23:824
102. Krishnamoorti R, Giannelis EP (1997) Macromolecules 30:4097
103. Heinrich G, Klueppel M (2002) Adv Polym Sci 160:1

104. Costa FR, Wagenknecht U, Jehnichen D, Goad MA, Heinrich G (2006) Polymer 47:1649
105. Subbotin A, Semenov A, Manias E, Hadziioannou G, Brinke GT (1996) Macromolecules 28:1511
106. Solomon MJ, Almusallam AS, Seefeldt KF, Somwangthanaroj A, Vardan P (2001) Macromolecules 34:1864
107. Krishnamoorti R, Ren J, Silva AS (2001) Macromolecules 114:4968
108. Stratton RA, Butcher AF (1973) J Polym Sci: Polym Phys Ed 11:1747
109. Doi M, Edwards SF (1979) J Chem Soc Farad Trans 2: Mol Chem Phys 75:38
110. Doi M, Edwards SF (1986) The theory of polymer dynamics. University Press, Oxford
111. Doi M, Edwards SF (1978) J Chem Soc Farad Trans 2: Mol Chem Phys 74:1802
112. Wagner MH (1976) Rheol Acta 15:136
113. Osaki K (1993) Rheol Acta 32:429
114. McLeish TCB, Larson RG (1998) J Rheol 42:81
115. Lertwimolnun W, Vergnes B, Ausias G, Carreau PJ (2007) J Non-Newt Fluid Mech 141:167
116. Ren J, Krishnamoorti R (2003) Macromolecules 36:4443
117. Larson RG (1984) J Rheol 28:545
118. Wagner MH, Schaeffer J (1992) Rheol Acta 36:1
119. Wagner MH, Yamaguchi M, Takahashi M (2003) J Rheo 47:779
120. Verbeeten WMH, Peters GWM, Baaijens FPT (2007) J Non-Newt Fluid Mech 117:73
121. Bird RB, Armstrong RC, Hassager O (1987) Dynamics of polymeric fluids. Wiley, New York
122. Saphiannikova M, Costa FR, Wagenknecht U, Heinrich G (2007) Polym Sci A (in press)
123. Mujumdar A, Beris AN, Metzner AB (2002) J Non-Newt Fluid Mech 102:157
124. Li J, Zhou C, Wang G, Zhao D (2003) J Appl Polym Sci 89:3609
125. Gilman JW, Morgan AB, Harris R Jr, Manias E, Giannelis EP, Wuthenow M (1999) Proceedings fire retardant chemicals association, Tuscon, AZ, 24–27 October 1999. FRCA, Lancaster PA, USA, p 9
126. Huggett C (1980) Fire Mater 4:61
127. Wang Z, Qu B, Fan W, Huang P (2001) J Appl Polym Sci 81:206
128. Scudamore MJ, Briggs PJ, Prager FH (1991) Fire Mater 15:65
129. Bartholmai M, Schartel B (2004) Polym Advan Technol 15:355
130. Schartel B, Poetschke P, Knoll U, Abdel-Goad M (2005) Eur Polym J 41:1061
131. Wagenknecht U, Kretzschmar B, Reinhardt G (2003) Macromol Symp 194:207
132. Zammarano M, Franceschib M, Bellayera S, Gilmana JW, Meriani S (2005) Polymer 46:9314
133. Beyer G (2001) Fire Mater 25:193
134. Hsveh HB, Chen CY (2003) Polymer 44(4):1151–1161
135. Costa FR, Wagenknecht U, Heinrich G (2007) Polym Degrad Stab 92(10):1813–1823

Editor: Karel Dušek

Adv Polym Sci (2008) 210: 169–208
DOI 10.1007/12_2007_120

Published online: 11 October 2007

Synthesis of Stimuli-Responsive Polymers by Living Polymerization: Poly(*N*-Isopropylacrylamide) and Poly(Vinyl Ether)s

Sadahito Aoshima (✉) · Shokyoku Kanaoka

Department of Macromolecular Science, Graduate School of Science,
Osaka University, Toyonaka, 560-0043 Osaka, Japan
aoshima@chem.sci.osaka-u.ac.jp

Abstract Precision synthesis via living polymerization has created new possibilities to a variety of stimuli-responsive polymers. This review highlighted the synthesis of poly(*N*-

isopropylacrylamide) [poly(NIPAM)] and poly(vinyl ether)s by living polymerization. Poly(NIPAM) is a well-known thermoresponsive polymer, and the related well-defined polymers including block copolymers and end-functionalized polymers have been synthesized intensively throughout the world since the discovery of its living radical polymerization. This recent revolutionary change in the investigations on poly(NIPAM) is first described. Poly(vinyl ether)s can be stimuli-responsive with high sensitivity when having certain functional groups. A variety of stimuli-responsive polymers with controlled sequences and/or shapes such as block, gradient copolymers, and star-shaped polymers were designed and synthesized by living cationic polymerization in the presence of an added base. The self-assembling behavior of the obtained polymers is also demonstrated. In addition to the selective synthesis of stimuli-responsive polymers, the recent development of initiating systems for the vinyl ether living polymerization by our group is briefly reviewed.

Keywords Block copolymers · Living polymerization · Self-assembly · Stimuli-responsive polymers · Thermoresponsive

Abbreviations

AIBN	2,2′-azobis(isobutyronitrile)
ATRP	Atom transfer radical polymerization
AzoVE	4-[2-(vinyloxy)ethoxy]azobenzene
BMDO	5,6-benzo-2-methylene-1,3-dioxepane
DMF	*N*,*N*-dimethylformamide
DSC	Differential scanning calorimetry
EOEOVE	2-(2-ethoxy)ethoxyethyl vinyl ether
EOVE	2-ethoxyethyl vinyl ether
GPC	Gel permeation chromatography
IBEA	1-isobutoxyethyl acetate [$CH_3CH(OCOCH_3)OCH_2CH(CH_3)_2$]
IBVE	Isobutyl vinyl ether
LCST	Lower critical solution temperature
Me_6TREN	tris(2-dimethylaminoethyl)amine
MOVE	2-methoxyethyl vinyl ether
MPC	2-methacryloyloxyethyl phosphorylcholine
MWD	Molecular weight distribution
NIPAM	*N*-isopropylacrylamide
NMP	Nitroxide-mediated polymerization
NMR	Nuclear magnetic resonance
ODVE	Octadecyl vinyl ether
PBO	Poly(butylene oxide)
PBS	Phosphoric acid buffer solution
PEO	Poly(ethylene oxide)
PLA	Poly(lactic acid)
PNIPAM	Poly(*N*-isopropylacrylamide)
PPO	Poly(propylene oxide)
PS	Polystyrene
RAFT	Reversible-fragmentation chain transfer
TEM	Transmission electron micrography
TEMPO	2,2,6,6-tetramethylpiperidine 1-oxyl
T_{PS}	Phase separation temperature

UCST	Upper critical solution temperature
UV	Ultraviolet

1 Introduction

Stimuli-responsive polymers whose properties and shape change in response to an external stimulus have recently been attracting attention [1, 2]. Moreover, the scope of the investigations in this area ranges from academic research to the applications of these materials in various fields including polymer design/synthesis, physical chemistry (e.g., changes in morphology, self-assembly), smart gels, and biomedical disciplines [3–7]. These polymers are therefore also called "smart", "intelligent", or "environmentally sensitive" polymers.

Sensitivity, reversibility, accuracy, and self-assembly of such polymers are the keys to constructing "intelligent" stimuli-responsive systems and would be affected by the polymer primary structures. However, systematic investigations of the relationships between polymer structures and stimuli-responsive behavior were limited until several years ago, as the living polymerizations of related monomers involved in these syntheses were difficult to carry out. For example, the structure and molecular weight of polymers could not be freely controlled either for the conventionally investigated thermoresponsive polymers such as cellulose derivatives [8], poly(ethylene oxide) (PEO) derivatives [9], poly(methyl vinyl ether) [10], partly hydrolyzed poly(vinyl acetate) [11], and poly(*N*-alkylacrylamide)s [12, 13], or for poly(*N*-isopropylacrylamide) (PNIPAM), on which there have been advanced studies [14–18].

Living polymerization has recently been extended to radical polymerization [19–21], complementing existing living anionic, cationic, and coordination polymerizations. This breakthrough has expanded possibilities for the precision synthesis of various stimuli-responsive polymers because living radical polymerization can be achieved with a range of commercial functional monomers. There are a number of advances stemming from progress in living polymerization methods: (i) a series of polymers having a high response sensitivity have been successfully designed/synthesized, (ii) a method has been discovered for introducing a functional group to a specific position of the polymer, (iii) the synthesis of block/graft copolymers with well-defined molecular weight and structure has become possible. In other words, the conventional synthetic difficulties have diminished and systematic studies on stimuli-responsive polymers have resumed.

In this paper, the living radical polymerization of *N*-isopropylacrylamide (NIPAM) and the living cationic polymerization of vinyl ethers having oxyethylene groups are outlined (Fig. 1), as they constitute a new wave of the

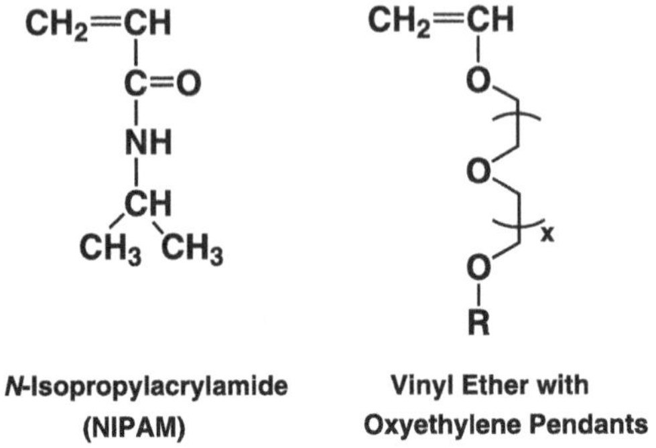

Fig. 1 Structures of *N*-isopropylacrylamide (NIPAM) and vinyl ethers with oxyethylene pendants

synthesis of thermoresponsive polymers (Scheme 1). In the former case, the main focus of this discussion is how the NIPAM system, the study of which has been undertaken throughout the world, has changed significantly in the last several years since the discovery of its living radical polymerization. The discussion of the latter case mainly concerns the results of our studies, and although poly(vinyl ether)s are minor polymers compared to PNIPAM, it is one of the early systems that demonstrates the usefulness and possibilities of living polymerization for systematic investigations on stimuli-responsive polymers. Here, in order to focus on the synthesis of such stimuli-responsive

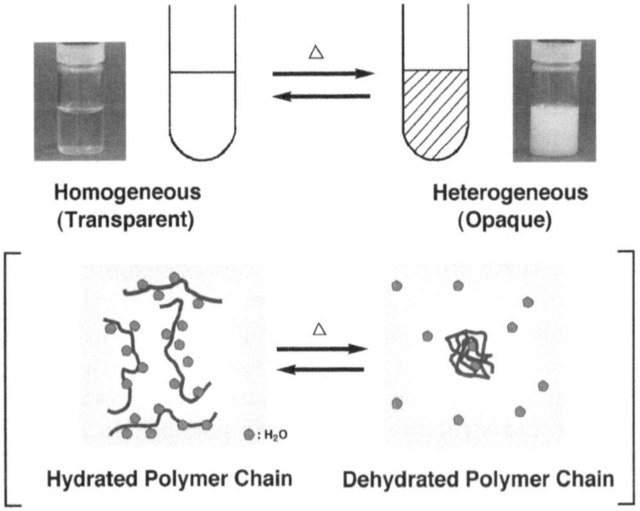

Scheme 1 Phase separation of thermoresponsive polymers in water [photographs: poly(2-ethoxyethyl vinyl ether) in water (200 mer, 1 wt % aq.)] and schematic illustration for the mechanism

polymers, the discussion of their physical chemistry and applications has been kept to a minimum. Interested readers are referred to other general reviews or books [22–25].

2 Living Polymerization of NIPAM

2.1 Development of Living Radical Polymerization and Radical Polymerization of NIPAM

Many methods have been developed for the polymerization of NIPAM at 50–70 °C in organic solvents using 2,2′-azobis(isobutyronitrile) (AIBN) or peroxide initiators, or in water using ammonium persulfate or potassium persulfate initiator in the presence of activators [14, 15]. Also, in order to control molecular weight and structure, especially for the functionalization of end-groups, polymerization in the presence of various chain transfer agents has been examined and applied to the synthesis of block copolymers. However, there has been a need for the synthesis of temperature-responsive polymers with more precisely controlled structure and molecular weight, and studies on the living polymerization of NIPAM have been carried out starting from about the year 2000.

Living radical polymerization [26] began with the synthesis of a styrene block copolymer with the "iniferter" method by Otsu et al. in 1982 [27]. During the 1990s, nitroxide-mediated polymerization (NMP) [28–30], atom transfer radical polymerization (ATRP) [31–33], and reversible addition-fragmentation chain transfer (RAFT) polymerization [34–37] were successively investigated; polymers having a narrow molecular weight distribution (MWD) and a variety of block copolymers were synthesized from various monomers. The study of the living polymerization of NIPAM was delayed in comparison with other monomers due to the various side reactions that occurred during its polymerization, and solubility problems with the polymer (leading to difficulties with GPC measurements). However, the living radical polymerization of NIPAM by the RAFT method has recently been demonstrated [35–42]. This method of polymerization used dithioesters as chain transfer agents, and could be applied to many polar monomers at a temperature lower than that of NMP and under milder polymerization conditions than those of ATRP [19–21, 37].

Numerous reports describing the successful RAFT polymerization of NIPAM have appeared [35]. For example, in 2000, Rizzardo et al. first reported the RAFT living radical polymerization of NIPAM by AIBN employing benzyl dithiobenzene or cumyl dithiobenzoate RAFT agents **1** (Fig. 2) [39]. Fukuda et al. synthesized block copolymers using these systems [40]. Subsequently,

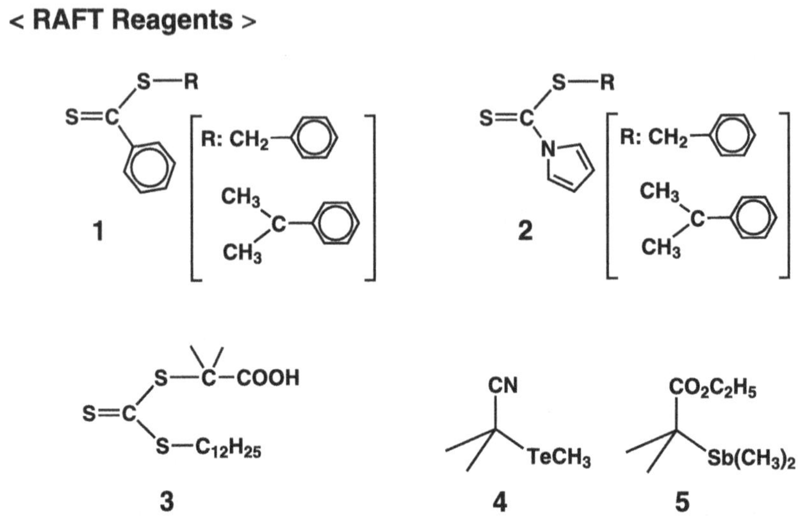

Fig. 2 Structures of RAFT reagents

Müller et al. achieved the benzyl and cumyl dithiocarbamate(**2**)-mediated polymerization of NIPAM [41]. McCormick et al. demonstrated controlled RAFT polymerization of NIPAM at room temperature using trithiocarbonate RAFT agent **3** in conjunction with an azo initiator in DMF or aqueous media, and succeeded in synthesizing a living polymer having an extremely narrow MWD [42]. Yamago et al. demonstrated the living polymerization of various monomers including NIPAM using an azo-initiator and organotellurium or organostibine compounds **4**, **5** as the chain transfer agents [43–45]. Other novel means, such as γ- [46] or UV-irradiation [47, 48], have been used instead of an azo-initiator to initiate the polymerizations.

Okamoto et al. demonstrated the simultaneous control of the stereostructure and molecular weight. The tacticity control in the RAFT polymerization of NIPAM was achieved by the addition of a suitable Lewis acid such as $Y(OTf)_3$ or $Sc(OTf)_3$; the synthesis of PNIPAM having a narrow MWD and an unusual tacticity ($m = 45–72\%$) became possible [49–52].

As for the other methods mentioned above, NMP and ATRP living polymerizations of NIPAM have also been reported very recently. NMP has been carried out with initiator **6** (Fig. 3), α-hydrogen alkoxyamine derivatives, instead of the TEMPO-based systems [53–55]. Although the living polymerization behavior has not been reported in detail, the synthesis of homopolymer of NIPAM with a narrow MWD, its block copolymers with polystyrene (PS) [53], and well-controlled star block copolymers having PNIPAM and PS segments was achieved by Hawker and Fréchet et al. [54]. On the other hand, the successful ATRP of NIPAM has been reported by Masci et al. [56] and Stöver et al. [57]. In each case, alkyl 2-chloropropionate (**7**), copper(I)

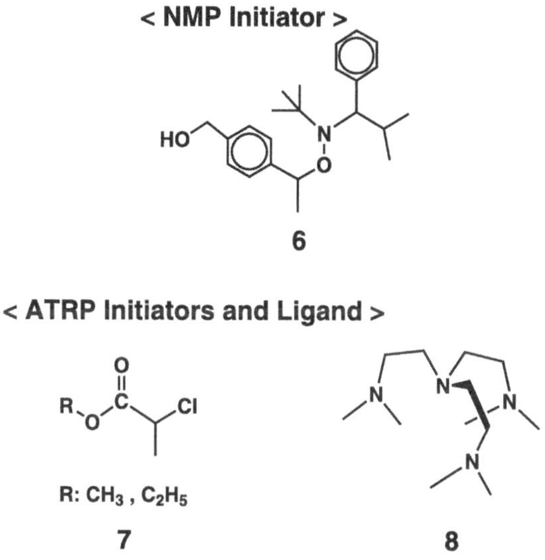

Fig. 3 Structures of NMP initiator, ATRP initiators, and ATRP ligand

chloride, and Me_6TREN (**8**) were used as initiator, catalyst, and ligand, respectively. The polymerizations were carried out in alcohols or DMF/water at around room temperature to give well-controlled PNIPAM with a narrow MWD.

The effect of molecular weight of PNIPAM has conventionally been examined by complicated fractionation [16] or other methods. As a result of these advances, it has become possible to examine the influence of molecular weight on the lower critical solution temperature (LCST) of aqueous solutions of PNIPAM [57–59]. No decisive conclusions have yet been obtained, yet detailed investigation by several groups has started including the effect of end groups.

2.2
Living Anionic Polymerization of NIPAM

Many studies have been carried out on the anionic polymerization of NIPAM using conventional organometallic catalysts. However, the acidic amide proton of NIPAM inhibited anionic polymerization, instead inducing hydrogen-transfer polymerization under basic conditions, as shown in Scheme 2. The polymers obtained were insoluble in water and did not show LCST phase separation [14, 15]. Recently, the polymerization of monomers in which this active hydrogen was protected by methoxymethyl or trialkylsilyl group **9**, **10** was examined and living anionic polymerization has become possible [60–62]. Ishizone et al. have prepared atactic PNIPAM with a narrow MWD via the anionic polymerization of *N*-methoxymethyl-NIPAM (**9**)

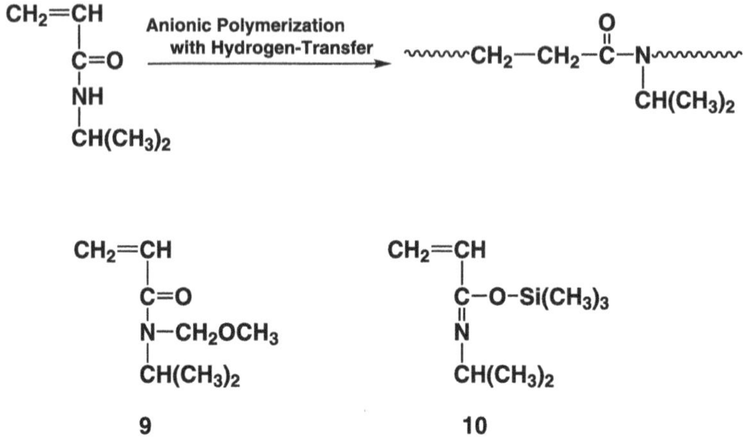

Scheme 2 Mechanism of conventional anionic polymerization of NIPAM with hydrogen-transfer and structures of NIPAM protected by methoxymethyl or trialkylsilyl group

with diphenylmethylpotassium in the presence of Et_2Zn and subsequent acidic hydrolysis [60, 61]. The stereoregularity of PNIPAM could be changed over a wide range from isotactic ($m = 85\%$) to syndiotactic ($r = 83\%$) by an appropriate choice of counterions for the anionic initiators or by the use of additives. Kitayama et al. have reported the synthesis of highly isotactic PNIPAM ($m = 80–97\%$) via anionic polymerization of trimethylsilyl-protected NIPAM (**10**) [62]. This isotactic PNIPAM had much lower solubility in water.

The anionic polymerization of NIPAM derivatives having an LCST has also been examined in detail [63]. Because these monomers had no acidic amide protons, the living polymerization proceeded easily in comparison with NIPAM, and the influence of tacticity, etc., has been examined.

3
Synthesis of Various Functionalized NIPAM Polymers

The establishment of the living radical polymerization of NIPAM encouraged not only many polymer chemists but also polymer physicists to prepare functionalized NIPAM polymers with various controlled sequences and/or shapes, as discussed in this chapter. In the following parts, block, random, or graft copolymers will be simply designated by the acronym A–B for a diblock copolymer, A–B–A for an ABA-type triblock copolymer, A–B–C for an ABC-type triblock copolymer, A-*co*-B for a random copolymer, A-*g*-PNIPAM for grafting of NIPAM segments onto the polymer A. For example, PNIPAM-PEO stands for a diblock copolymer of PNIPAM and PEO.

3.1
Synthesis of Thermoresponsive Block Copolymers and End-Functionalized Polymers

Until the recent development of living radical polymerization, block copolymers of NIPAM were synthesized using chain transfer agents. For instance, Feijen et al. generated an end radical of PEO by the redox system using $Ce^{(IV)}$ to initiate the polymerization of NIPAM [64–67]. On the other hand, Okano et al. introduced amino or hydroxy groups at the end of PNIPAM by the "telomerization" method using aminoethanethiol or hydroxyethanethiol as telomers, respectively, to synthesize various block copolymers or comb-type grafted hydrogels [68–70].

Since 2000, living radical polymerization has allowed the synthesis of block copolymers with controlled molecular weight and a narrow MWD using many monomer combinations, as shown in Figs. 4 and 5 [71]. Block copolymerization with hydrophilic or thermoresponsive acrylamide derivatives **11, 12** has been examined [72–77]. Block copolymers having hydrophilic segments such as PNIPAM-poly[(meth)acrylic acid] (**13**) [78–80], PNIPAM-poly(sulfonic acid) (**14**) [81, 82], PNIPAM-poly(2-hydroxyethylacrylate) (**15**) [83], and PNIPAM-{poly[(2-dimethylamino)ethyl methacrylate]-*co*-poly(2-hydroxyethyl methacrylate)} (**16**) [84] were prepared. These formed polymer micelles in response to variation of the temperature. For example, Müller et al. have synthesized PNIPAM-poly(acrylic acid) with low polydis-

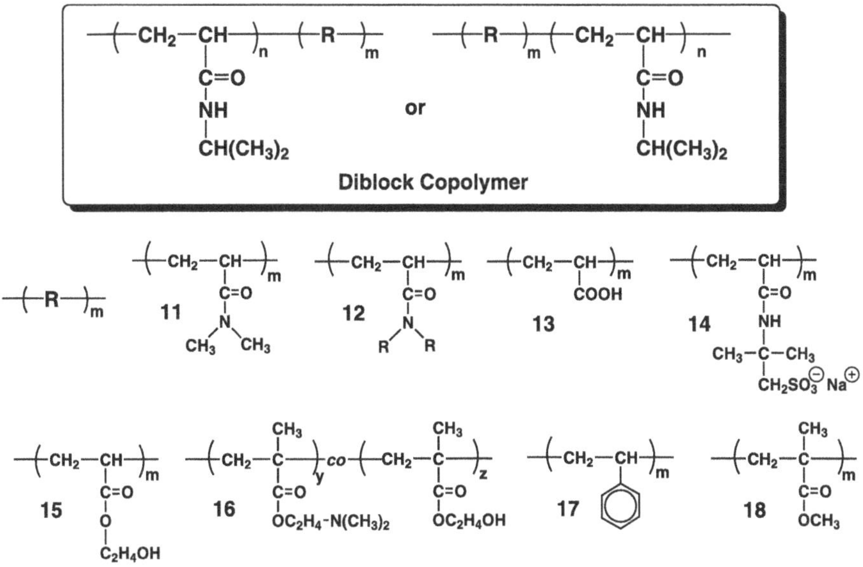

Fig. 4 Structures of diblock copolymers of NIPAM (**1**)

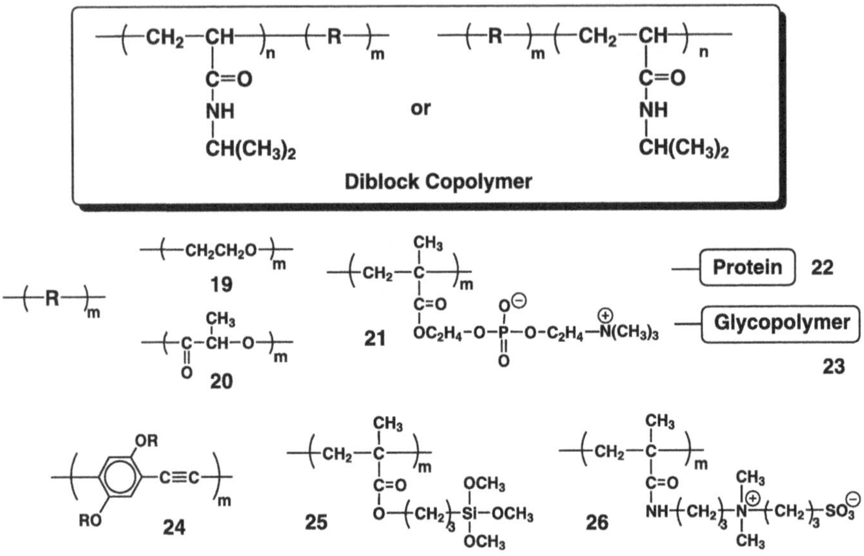

Fig. 5 Structures of diblock copolymers of NIPAM (2)

persities by RAFT polymerization. The block copolymers formed micelles in aqueous solution in dependence of temperature and pH, which was investigated by DLS, temperature-sweep NMR, IR, and cryo-TEM [78]. On the other hand, block copolymers with hydrophobic segments including PS (**17**) [53, 85, 86] or poly(methyl methacrylate) (**18**) [87] have also been synthesized. For instance, PNIPAM-PS was synthesized by NMP using α-hydrogen alkoxyamine derivatives [53] or by RAFT polymerization [85, 86].

The synthesis of biodegradable (or biocompatible) and thermoresponsive block copolymers with PEO (**19**) [88–91], poly(lactic acid)s (PLA) (**20**) [92, 93], and poly(2-methacryloyloxyethyl phosphorylcholine) (PMPC: **21**) [94] has been achieved. In the last case, biocompatible and thermoresponsive ABA triblock copolymers (PNIPAM-PMPC-PNIPAM) were synthesized using ATRP by Armes et al. In aqueous solution, these triblock copolymers form free-standing physical gels at 37 °C reversibly [94]. To prepare PNIPAM-protein conjugates with retention of bioactivity, block copolymers of PNIPAM and protein (streptavidin, bovine serum albumin) (**22**) [95–100] or glycopolymer (**23**) [101, 102] have been prepared. Stayton, Hoffman, and Müller et al. [95, 96] synthesized conjugates of streptavidin and PNIPAM (or PNIPAM-PAA) with low polydispersities by RAFT polymerization. The conjugates rapidly formed mesoscale polymer-protein particles above the LCST [95, 96]. Maynard et al. synthesized bioactive "smart" polymer conjugates by polymerizing (ATRP method) from defined initiation sites on proteins [97, 98]. Furthermore, several functionalized block copolymers containing water-soluble poly(phenylene-ethynylene) segments (**24**) for fluorescent semiconducting applications [103, 104], Si-containing segments for fabrication of hybrid nanoparticles (**25**) [105], and zwitterionic segments (**26**) that exhibit upper critical solution temperature (UCST) [106, 107] have been prepared. For example, Tenhu and Laschewsky et al. have synthesized water-soluble block copolymers with double thermoresponsivity from NIPAM and 3-[*N*-(3-methacrylamidopropyl)*N*,*N*-dimethyl]aminopropane sulfonate. Such block copolymers exhibited switching between the interior and exterior of aggregates [106, 107].

Various types of block copolymers such as ABA triblock [93, 94, 105, 108], ABC triblock [109–111], PNIPAM-dendrimer [112], and PNIPAM-[star-shaped (or hyperbranched) polymer] [65], and PNIPAM-polyrotaxane-PNIPAM [66] have also been designed/synthesized. For example, Armes et al. have synthesized biocompatible, thermoresponsive ABC and ABA triblock copolymers such as PPO-PMPC-PNIPAM via ATRP using a PPO-based macroinitiator. McCormick et al. have reported the RAFT synthesis of thermally responsive ABC triblock copolymers incorporating *N*-acryloxysuccinimide for facile in situ formation of micelles with cross-linked shells.

Also, various end functional groups have been introduced by transformation of RAFT polymer end groups or by the use of functionalized initiators [113–116], and the influence of these end groups has been examined in detail [115]. Furthermore, the following groups were attached to one or both ends of PNIPAM, as shown in Fig. 6: pyrenyl groups (**27**) for fluorescent

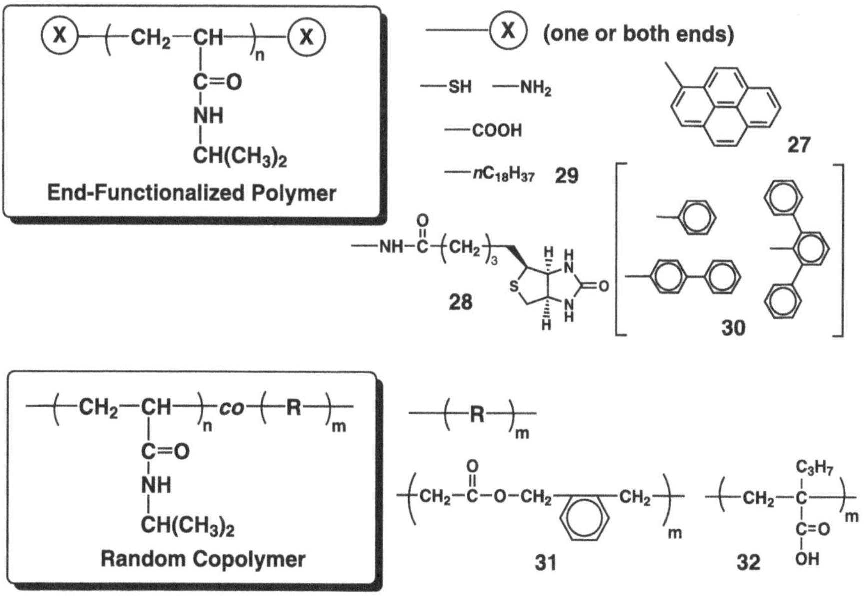

Fig. 6 Structures of end-functionalized PNIPAMs and random copolymers of NIPAM

labeling [117, 118], biotinyl groups (**28**) for potential applications in biological systems [34, 39, 65, 119], strongly hydrophobic alkyl groups containing octadecyl (**29**) [120–122] or benzene derivatives (**30**) [123], and bipyridyl groups (to the center of PNIPAM) for the synthesis of ruthenium-centered metallopolymers [124].

Several characteristic random copolymers were also prepared, as shown in Fig. 6 [125–127]. Matyjaszewski et al. have reported the development of an injectable thermoresponsive hydrogel [poly(NIPAM-*co*-BMDO)] (**31**) consisting of PNIPAM with degradable units as an injectable scaffold to enhance fracture repair. Poly(NIPAM-*co*-BMDO) was prepared by ATRP and RAFT [125]. Temperature- and pH-sensitive random copolymers of NIPAM and propylacrylic acid (**32**) were prepared using RAFT polymerization by Stayton and Hoffman et al. The dual temperature and pH responses were characterized, and their sharp and tunable phase transitions were demonstrated around neutral pH [126].

3.2 Synthesis of NIPAM Polymers with Various Shapes

Dendrimers of various shapes with NIPAM segments were synthesized by the RAFT method for nanomaterial applications, including drug release, diagnostics, catalysis, and separations [53, 128–133]. For example, You and Pan et al. have prepared dendritic core-shell nanostructures with a PNIPAM shell using dendritic RAFT agents with many terminal dithioesters on the dendrimer surface. The aqueous solution showed reversible changes in turbidity, becoming transparent below and opaque above the LCST. Amphiphilic dendritic-linear diblock copolymers have also been synthesized. TEM results indicated the presence of spherical micelles with thermoresponsive shells in aqueous solution [128]. Kono et al. prepared poly(amidoamine) or poly(propyleneimine) dendrimers having isobutyl amide (IBAM) groups at all the chain ends. The attachment of IBAM, of which structure is similar to that of NIPAM, could change the nature of the dendrimer surface between hydrophilic to hydrophobic in response to the temperature variation, which induced characteristic thermosensitive phase separation [133]. Further, various types of branched polymers have been designed and prepared, such as star-shaped polymers [54, 65, 134–139], hyperbranched polymers [140–144], and miktoarm star polymers [145]. Fréchet and Hawker et al. have explored new strategies for the preparation of functional, multiarm star polymers via NMP [54].

In other endeavors to prepare NIPAM nanoparticles, the introduction of NIPAM brushes onto fine particles [146] or NIPAM nanotubes [147] has been attempted. Furthermore, Kawaguchi et al. have prepared various types of thermoresponsive "hair" particles by living radical polymerization using the "photoiniferter" method [148–150]. These particles demonstrated certain unique behaviors in terms of electrophoretic mobility and adsorption of dye molecules, as well as swelling/deswelling, and self-assembled into a two-dimensional superlattice when their dilute dispersions were dried on substrates [149]. Lu and Ballauff et al. synthesized monodisperse thermoresponsive PS-PNIPAM core-shell particles composed of a PS core and a cross-linked PNIPAM shell by UV-induced photoemulsion (radical) polymerization [151].

3.3 Grafting of NIPAM Segments onto Various Polymers or Inorganic Substrates

Living radical polymerization has been used to graft a thermosensitive segment onto various polymers such as polysaccharides (**33**) [152], dextran (**34**) [153], peptide nanotubes (**35**) [154], and several polymers including PS (**36**) [155], polyesters (**37**) [156], polypropylene (**38**) [157], poly(vinylidene fluoride) (**39**) [158], poly(hydroxymethacrylate) (**40**) [159], PNIPAM gel (**41**) [160–162] (Fig. 7). The graft efficiency was very high in each case because of the use of living radical polymerization. For example, a versatile ATRP method for grafting polysaccharides under mild homogeneous conditions has been reported by Masci et al. [152]. Schmidt et al. have reported on the synthesis of cylindrical-brush polymers comprising PNIPAM

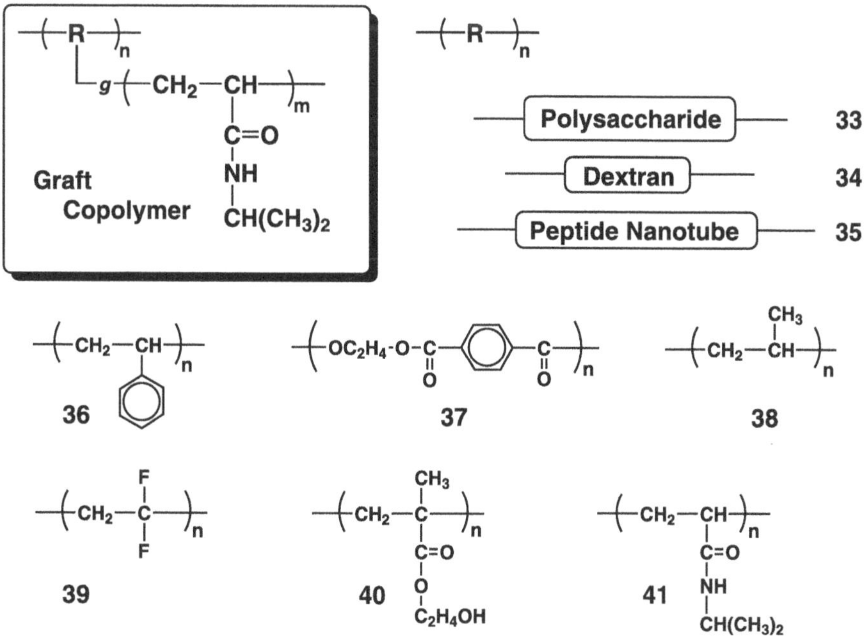

Fig. 7 Structures of graft copolymers with PNIPAM

side chains. In aqueous solution, the side chain repulsion could be controlled easily by variation of the temperature [159]. Lu et al. have synthesized comb-type grafted hydrogels, which comprises PNIPAM (or hydrophilic) backbone and freely mobile PNIPAM graft chain, by RAFT polymerization and end-linking processes. The gels exhibited rapid swelling/deswelling kinetics [160–162].

Living radical polymerization was also very effective in grafting these polymers onto inorganic substrates such as gold surface [163–172], gold nanoparticle [173–175], silica surface [176, 177], Si(100) [178–180], aluminum oxide membrane [181], and glass capillary [182], and for potential use in lithography [183], etc. For example, thermoresponsive gold nanoparticles or clusters and well-defined NIPAM brushes on gold membranes were prepared via surface-initiated ATRP or by use of end-functionalized PNIPAM with a thiol group. Further, studies on water-soluble multiwalled carbon nanotubes with grafted PNIPAM shells [184–187] and grafts on carbon blacks have also been reported [188]. Recently, a nano-composite NIPAM gel was synthesized using radical polymerization in the presence of a clay mineral by Haraguchi et al. Although it was not a living polymerization, the resulting gels exhibited extraordinary mechanical, optical, and swelling/deswelling properties [189].

4 Synthesis of Other Thermoresponsive Polymers

4.1 PEO-Related Block Copolymers

Thermoresponsive polymers other than PNIPAM were synthesized by living polymerization, particularly PEO-related triblock copolymers (Pluronic-type) due to their many commercial applications as emulsifiers, dispersants, and stabilizers [71, 190, 191]. The phase behavior of many PEO-related block copolymers has been studied by Wanka et al. and Alexandridis et al. [192, 193]. These were synthesized by anionic ring opening polymerization, PEO-poly(propylene oxide)-PEO (PEO–PPO–PEO: **42**) [194], PEO–PLA–PEO (**43**) [195], PEO–PPO (**44**) [190], PEO-poly(butylene oxide) (PEO–PBO: **45**) [196] being typical examples (Fig. 8). In all cases, because the hydrophobic and hydrophilic properties of each segment depended on temperature, the aggregation behavior changed with temperature at low concentration, such as unimers, spherical micelles, cylindrical micelles, and clouding. At high concentrations, a transparent gel with densely filled spherical micelles was generated [197]. Gelation of PEO–PPO–PEO triblocks occurred due to close packing of micellar aggregates above a certain critical concentration. In contrast, with reverse triblock copolymers (PPO–PEO–PPO), the PPO end-blocks

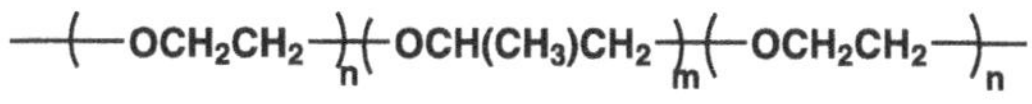

PEO-PPO-PEO (42)

PEO-PLA-PEO (43)

PEO-PPO (44)

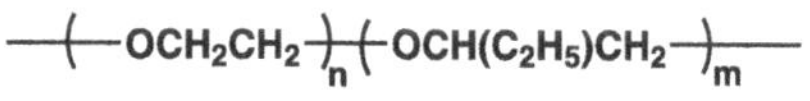

PEO-PBO (45)

Fig. 8 Structures of thermoresponsive PEO-related block copolymers

could bridge adjacent micelles, leading to the formation of a physical network [90, 198]. The research area of such PEO-related block copolymers has been reviewed by other authors [190, 199, 200].

4.2 Various Thermoresponsive Polymers

Polymethacrylate having oxyethylene groups on the side chains have been examined [201–203] in addition to the poly(vinyl ether)s that is described in the latter part of this article. The synthesis of water-soluble polymethacrylates was recently investigated via living anionic polymerization of oligo(ethylene glycol) methacrylates (**46**) with 1,1-diphenyl-3-methylpentyllithium/LiCl or diphenylmethylpotassium/$ZnEt_2$ initiator systems in THF at −78 °C. Aqueous solutions of polymethacrylates having two or three oxyethylene units showed reversible cloud points, and the surfaces of their block copolymers with polystyrene were characterized [201]. Watanabe et al. have synthesized various temperature-responsive polymers having an ether bond in the main and side chains (**47**) by anionic ring-opening polymerization [202], and have examined the temperature-dependent sol–gel transitions of the resultant block copolymers. Thermoresponsive poly(organophosphazene)s having oxyethylene and hydrophobic ester side groups (**48**) were synthesized by Sohn and Song et al., and their reversible sol–gel transition were investigated by ^{31}P NMR and viscometry [203] (Fig. 9).

Various thermoresponsive PNIPAM derivatives were also obtained by radical or ionic polymerization, in which most polymerizations were nonliving systems except for a few examples [63, 204], including poly(*N*-alkylacrylamide)s (**49**) [12, 13, 63], poly(2-isopropyl-2-oxazoline) (**50**) [204], poly(*N*-vinylisobutylamide) (**51**) [205], optically active poly(methacrylamide)

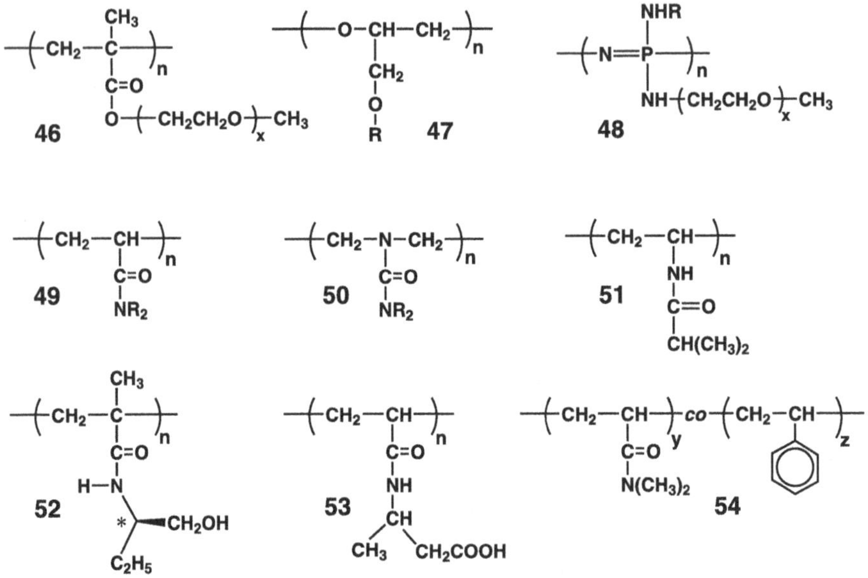

Fig. 9 Structures of thermoresponsive polymers

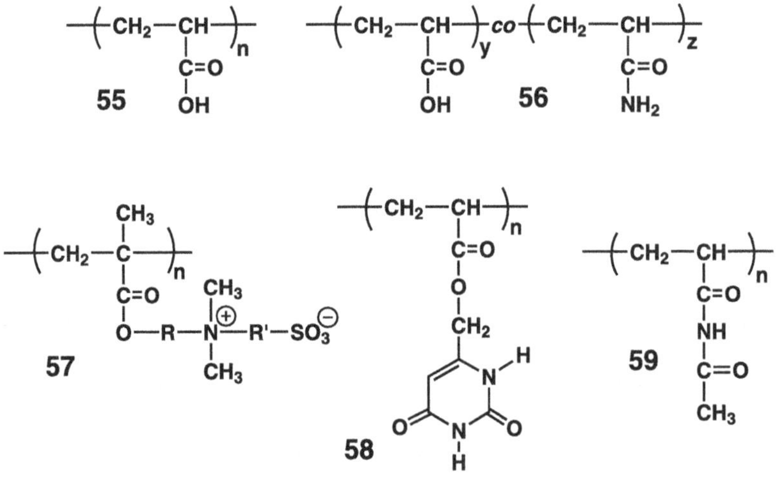

Fig. 10 Structures of thermoresponsive polymers exhibiting UCST phase separation in aqueous solution

(**52**) [206], poly(2-carboxy-NIPAM) (**53**) [207], and random copolymers of *N*,*N*-dimethylacrylamide and styrene (**54**) [208].

On the other hand, several polymers of acrylic acid (**55**), acrylamide (**56**), and (meth)acrylate with sulfobetain (**57**) or uracil side group (**58**) which undergo UCST (not LCST) phase separation, have been studied [209–214], as shown in Fig. 10. For example, Onishi et al. have prepared random copolymers containing *N*-acetylacrylamide (**59**), which exhibited UCST phase separation in aqueous solution, for preparing temperature-responsive magnetic nanoparticles [213, 214].

5 Stimuli-Responsive Poly(Vinyl Ether)s

Living cationic polymerization has led to precision synthesis of functionalized polymers from species such as vinyl ethers, isobutene, and styrene derivatives [215]. Among those, vinyl ether polymers allow a wide choice of functional groups, which, along with the possibility of precision synthesis, has led to the development of a variety of stimuli-responsive polymers with well-defined structures [216]. This chapter describes our recent study on the synthesis and stimuli-responsive behavior of various poly(vinyl ether)s obtained by living cationic polymerization in the presence of an added base.

5.1 Thermoresponsive Polymers

The thermoresponsive properties of poly(vinyl ether)s with oxyethylene pendants (**60**) in water were discovered about 20 years ago [217–219]. This phase transition was not only highly sensitive but also completely reversible on heating and cooling, without hysteresis. Clear aqueous solutions of these polymers rapidly transformed into turbid solutions, with complete transition within 1 °C. This was also the case for the reverse transition. For example, an aqueous solution of a polymer (**60a**) (Fig. 11) obtained from 2-ethoxyethyl vinyl ether (EOVE) underwent phase separation sensitively at 20 °C. A systematic survey revealed that the phase separation temperature (T_{PS}) of poly(vinyl ether)s with oxyethylene pendants can be controlled by varying

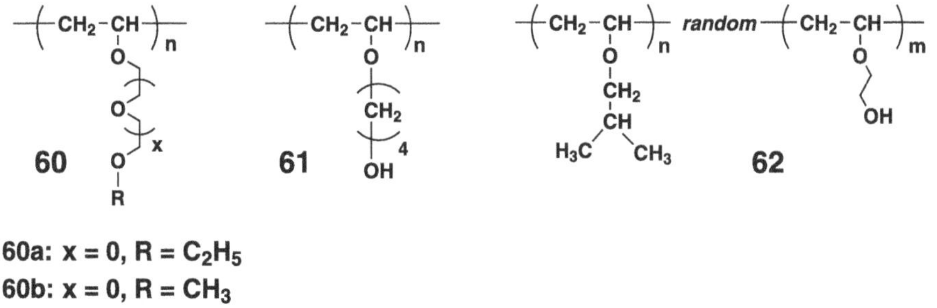

60a: x = 0, R = C_2H_5
60b: x = 0, R = CH_3
60c: x = 1, R = C_2H_5

Fig. 11 Structures of thermoresponsive poly(vinyl ether)s

the length of the pendant oxyethylene units and/or the hydrophobicity of an ω-alkyl group, or by regulating the composition of a random copolymer of two different thermosensitive units, the T_{PS} of which lies between those of the two homopolymers [218, 219]. It should also be noted that high sensitivity in phase transition was observed only with low-polydispersity polymers.

An appropriate hydrophilic/hydrophobic balance is the key to producing a polymer with thermosensitivity in water. This concept has been proven by our recent study demonstrating thermosensitivity of a polymer prepared from an alcohol-type monomer with the desired amphiphilicity [220]. A hydroxyalkyl group is a straightforward rearrangement of an oxyethylene unit, and is expected to have a similar overall amphiphilic balance, depending on the alkyl chain length. In addition, hydroxy groups are important for polymer self-assembly because of their hydrogen-bonding interactions. An aqueous solution of **61**, derived from a silyloxy-protected pendant counterpart, underwent rapid phase separation at 42 °C. The study also revealed that a fine balance between hydrophilicity and hydrophobicity was required for achieving thermosensitivity. For polyalcohols with a linear spacer between two oxygens in the side group, only the polymer with a $-(CH_2)_4-$ spacer exhibited reversible thermally induced phase separation with high sensitivity, whereas those with shorter or longer spacers were soluble and insoluble in water, respectively.

Since the pendant design is often cumbersome, the synthesis of more generalized thermoresponsive polymers was the next target. An appropriate hydrophilic/hydrophobic balance may be achieved in a random copolymer of hydrophobic and hydrophilic units. Thermoresponsiveness has been observed for a number of copolymers containing hydrophilic and hydrophobic units produced by free-radical polymerization. For these polymers, however, transitions are relatively slow and are accompanied by hysteresis between the heating and cooling cycles resulting from ill-defined structures. To avoid these disadvantages, we prepared random copolymers (**62**) containing hydrophilic units with pendant hydroxy groups and hydrophobic alkyl vinyl ether units via living cationic polymerization in the presence of an added base [221]. An aqueous solution of the product random copolymer was found to undergo sensitive and reversible thermally induced phase separation. Randomness of sequence distribution is essential for realization of such highly sensitive phase separation. For example, copolymers containing both block and random segments exhibited less sensitive phase separation behavior with hysteresis. The critical temperature of the random copolymer can be controlled by varying the composition of a hydrophobic vinyl ether and alcohol-containing units, and/or the structure of the hydrophobic repeating unit.

Most poly(vinyl ether)s are soluble in various organic solvents; hence, phase separation behavior in organic solvents is of interest. Thus, UCST-type phase separation was examined with various poly(vinyl ether)s. The driving forces for aggregation are divided into two types: (i) weak van der Waals interactions, and (ii) strong hydrophobic interactions, such as crystallization of long alkyl chains.

(i) Weak van-der-Waals Interaction

Linear alkane solutions of **60a** (n = 300, $M_n = 3.1 \times 10^4$, $M_w/M_n = 1.15$) showed highly sensitive UCST-type phase separation irrespective of the solvent [222]. Interestingly, the cloud point temperature of **60a** increased linearly with the number of carbon atoms in the alkane, which is in reasonable agreement with the Flory–Huggins theory. Similar phase separation occurred for poly(vinyl ether)s with various pendant groups, such as alkyl (in alcohols and esters), ester (in alcohols and toluene), and silyloxy groups (in alcohols). The combination of polymer and solvent was the decisive factor in sensitive phase separation. Nonpolar polymers underwent phase separation in polar solvent, and polar ones became thermosensitive in nonpolar media.

(ii) Strong Hydrophobic Interaction

Poly(octadecyl vinyl ether) [poly(ODVE)] (C18 pendant) with a very narrow MWD ($M_n = 3.0 \times 10^4$, $M_w/M_n = 1.07$) was soluble in alkanes, toluene, diethyl ether, THF, dichloromethane, and chloroform at room temperature but insoluble in acetone, ethyl acetate, and alcohols. Interestingly, sensitive UCST-type phase separation occurred for all solutions in various solvents [222, 223]. For some polar solvents in which the homopolymer was insoluble, random copolymers of ODVE with a polar monomer achieved similar sensitive UCST-type phase separation. A calorimetric study indicated that the transition with poly(ODVE) involves crystallization of pendant octadecyl chains. For a poly(ODVE) solution (10 wt %) in dichloromethane, for example, an exothermic peak in the DSC diagram due to crystallization was observed at about the critical temperature, determined from UV spectra of transmittance at 500 nm. On the other hand, no peak appeared in the thermogram of an ethyl acetate solution of poly(dodecyl vinyl ether) (C12 side chain: $M_n = 1.5 \times 10^4$, $M_w/M_n = 1.09$) containing long but amorphous alkyl pendants, although similar sensitive phase separation occurred at 46 °C. Pendant cholesteryl or biphenyl groups in poly(vinyl ether)s also induced sensitive UCST-type phase separation [224].

5.2 Other Stimuli-Responsive Polymers

Base-stabilized living cationic polymerization is tolerant towards various polar functional monomers, selectively yielding a new series of well-defined stimuli-responsive polymers including pH- [225], photo- [226–228], solvent- [229–231], and pressure-responsive [232] polymers (Scheme 3).

After the development of thermoresponsive polymers, our interest was directed towards pH-responsive behavior in polymer aqueous solutions, and

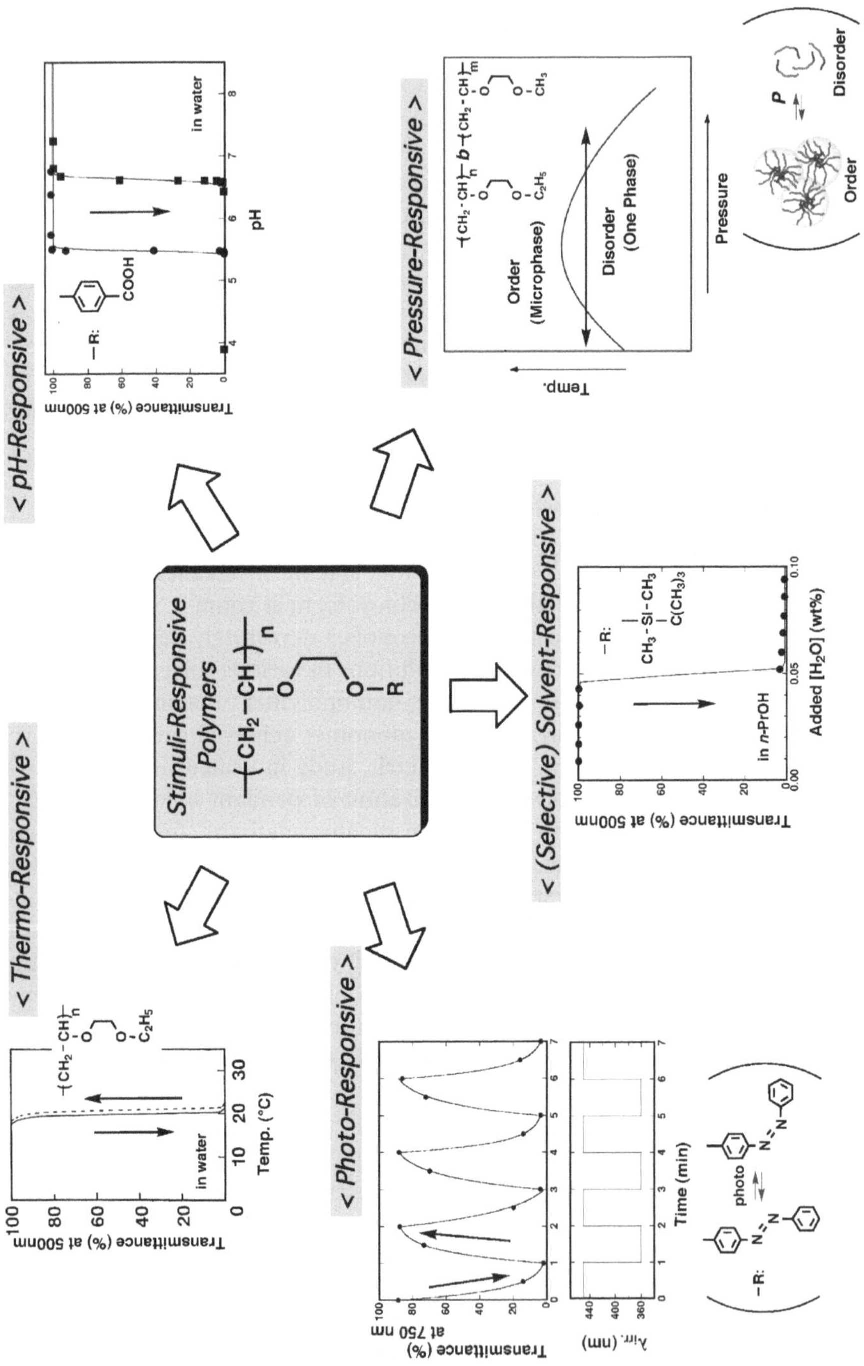

Scheme 3 Typical examples of various stimuli-responsive phase separation using poly(vinyl ether) derivatives

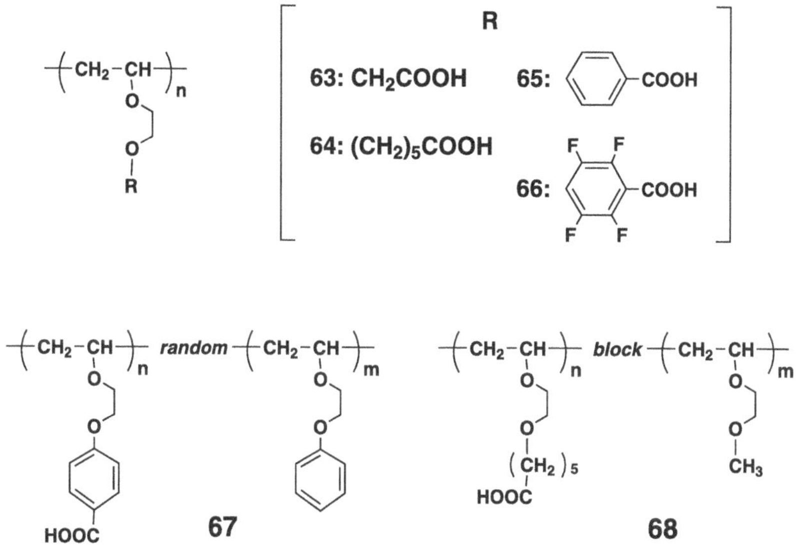

Fig. 12 Structures of pH-responsive poly(vinyl ether)s

carboxy-containing polymers were synthesized (Fig. 12). A solution of **63** was transparent at any pH. Although protonated carboxy groups in the acidic region are less hydrophilic, they are not hydrophobic enough to induce phase separation. The presence of a suitably hydrophobic substituent, such as an alkylene group (≥C5) (**64**) or a phenylene (**65**) adjacent to the carboxy group induced pH-sensitive phase separation [225]. The transmittance of solutions of polymers **64** and **65** sharply decreased at pH 5.6 and 6.5, respectively, when pH was reduced from the alkaline region. An important factor that determines the critical pH is the polymer pK_a value, as well as the solubility of the polymer. The critical pH for phase separation shifted to the acidic region when electron-withdrawing atoms were placed near the pendant carboxy groups (**66**). To enhance sensitivity in the transition, random copolymers with always-hydrophobic units were prepared. For a random copolymer, the critical pH increased with hydrophobic content (**67**). An appropriate amount of hydrophobic units resulted in sensitive aggregation in PBS even at pH 7.0, which is unusually high for carboxy-containing polymers.

Thus, four important points may be considered in controlling the critical pH value: (1) the solubility of substituents in the side chain, (2) the pK_a value of the polymer, (3) randomness in the distribution of hydrophobic units, and (4) the degree of polymerization. These factors may be easily controlled by using living cationic polymerization.

Azobenzene is well known for *cis-trans* isomerization of the $-N=N-$ bond induced by light irradiation. Living cationic polymerization of 4-[2-(vinyloxy)ethoxy]azobenzene (AzoVE) or its derivatives was achieved using

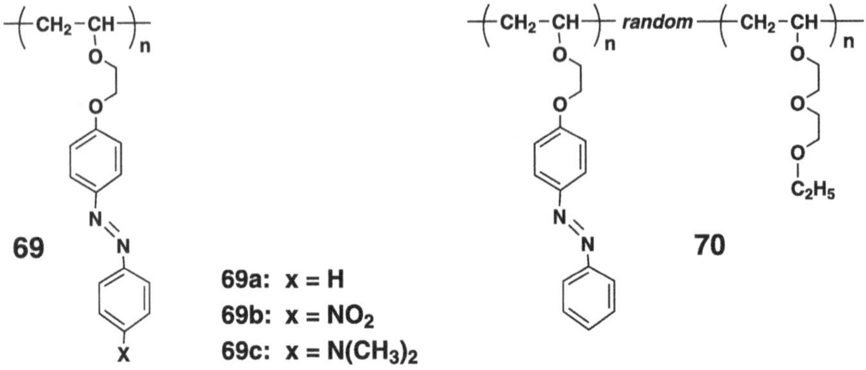

69

69a: x = H
69b: x = NO_2
69c: x = $N(CH_3)_2$

70

Fig. 13 Structures of photo-responsive poly(vinyl ether)s

various Lewis acids in the presence of an ester as an added base (Fig. 13) [226, 227]. Polymer **69a** was soluble in a diethyl ether/hexane mixture at 25 °C but became insoluble upon irradiation with UV light [226], under which most of the azo moieties are in the *cis* form. This phase transition behavior was sensitive and reversible upon irradiation with UV or visible light.

Random and block copolymers of AzoVE and various vinyl ethers (exhibiting LCST-type phase separation) were prepared ($[\mathrm{AzoVE}]_0 = 0.5–5$ mol %). For random copolymers (**70**) containing both thermally responsive and azobenzene units, phase separation occurred at a higher temperature under UV irradiation compared with that under visible light. This shift permitted solubility control of the polymer by irradiation with UV or visible light at a constant temperature [228].

Sensitive phase separation was also induced by addition of a small amount of an organic compound [229, 230] or water [229–231], or by increased pressure [232]. In the case of water-induced separation, rapid phase transition or physical gelation of a silyloxy-containing diblock copolymer was triggered by very small amounts of water (as little as 0.1%) [231].

5.3 Self-Association of Stimuli-Responsive Polymers with Controlled Sequences

After the discovery of thermosensitivity in poly(vinyl ether)s with oxyethylene side chains, these polymers were the only known living polymer systems that exhibited such sharp phase separation, except for PEO [190, 199, 200]. PNIPAM, known to behave in a similar manner, could not be obtained by living polymerization at the time; therefore, vinyl ethers with oxyethylene side groups were at an advantage, as they could be synthesized by living polymerization. Thus, we conducted a systematic investigation of vinyl ether systems, studying the effects of molecular weight, MWD, and other properties such as sequence distribution. For example, investigation of the effects of MWD showed, interestingly, that a narrow distribution corresponded to high sensitivity in phase separation [218]. This finding was quite important for our understanding of the system.

Living polymerization leads to the synthesis of well-defined multicomponent copolymers with a stimuli-responsive moiety, which usually aggregates in a phase-separated state. Based on the expectation that sharp and reversible phase separation might lead to construction of reversible self-association systems, thermally induced self-association of diblock copolymers was first examined [233–236]. A diblock copolymer (**71**) (Fig. 14) with a thermosensitive segment and a water-soluble segment was expected to form micelles in water above the critical temperature. Thus, diblock copolymers containing a thermosensitive segment with oxyethylene side chains and a polyalcohol segment were prepared by sequential living cationic polymerization [233,234]. These polymers, in fact, led to reversible micelles. The critical micelle concentration (CMC) and temperature (CMT) were determined based on solubilization behavior of a water-insoluble dye. The CMT was in good agreement with the clouding point of the thermosensitive seg-

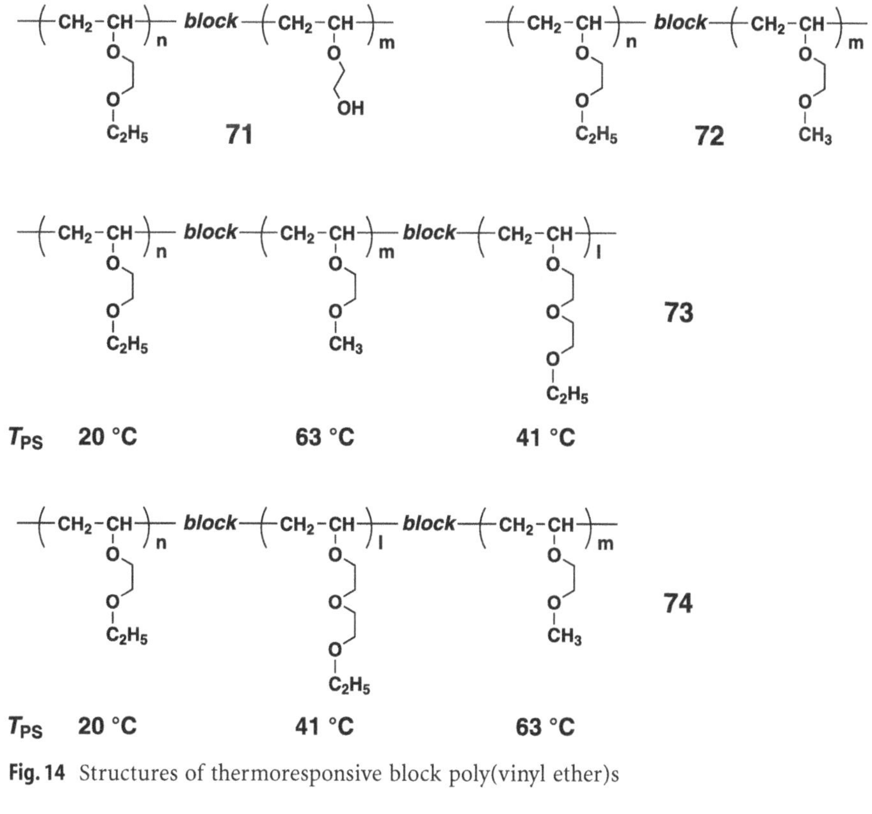

Fig. 14 Structures of thermoresponsive block poly(vinyl ether)s

ment [234, 237]. Interestingly, the resulting micelles had a very narrow size distribution.

At a higher concentration, the solution underwent rapid physical gelation upon warming to the critical temperature, giving a transparent gel, and reverted with sensitivity to the solution state at the same temperature upon cooling [233, 234]. For example, the flow behavior of a 20 wt % solution varied from Newtonian flow ($< 20\,^{\circ}$C) to non-Newtonian and plastic flow around the critical temperature within a very narrow temperature range. This was an unusual and puzzling result because diblock copolymers undergo physical crosslinking to form three-dimensional networks. A TEM picture of a freeze fracture replica of the obtained physical gel showed spherical aggregates densely packed across the field of view. Small-angle neutron scattering and dynamic light scattering studies revealed that the physical gel consisted of a micelle macrolattice with bcc symmetry [235, 236]. These results indicated that physical gelation starts with thermally induced micellization, giving nearly monodisperse spherical micelles with core size 18–20 nm. Subsequent and immediate macrolattice formation transforms the solution into a physical gel (Scheme 4) [234–236]. Based on the gelation mechanism, several stimuli-responsive gelation systems using diblock copolymers were created to respond to other stimuli such as addition of a selected solvent or compound [230–232]; cooling [222]; pH change [225]; pressure [232]; or irradiation with UV light [228], as shown in Scheme 5. The sharp transition of stimuli-responsive segments with highly controlled primary structure turned out to play an important role in self-association.

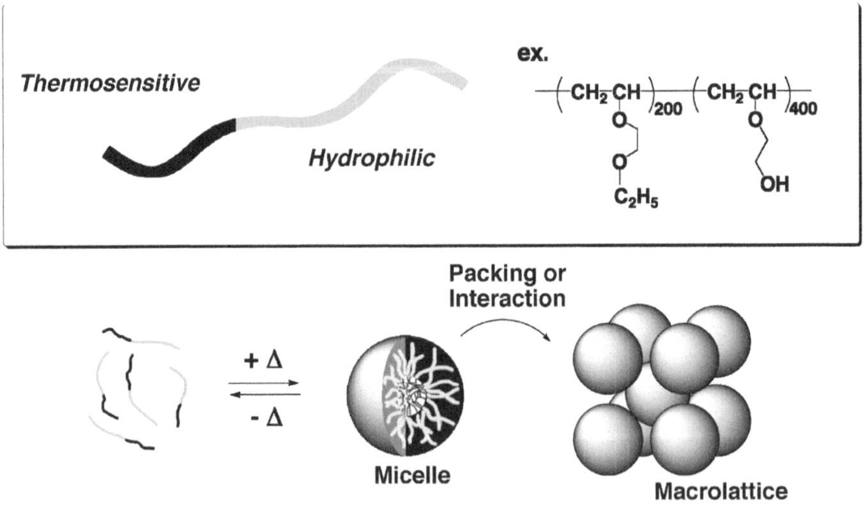

Scheme 4 Schematic illustrations for the mechanism of the physical gelation with thermoresponsive diblock poly(vinyl ethers)s

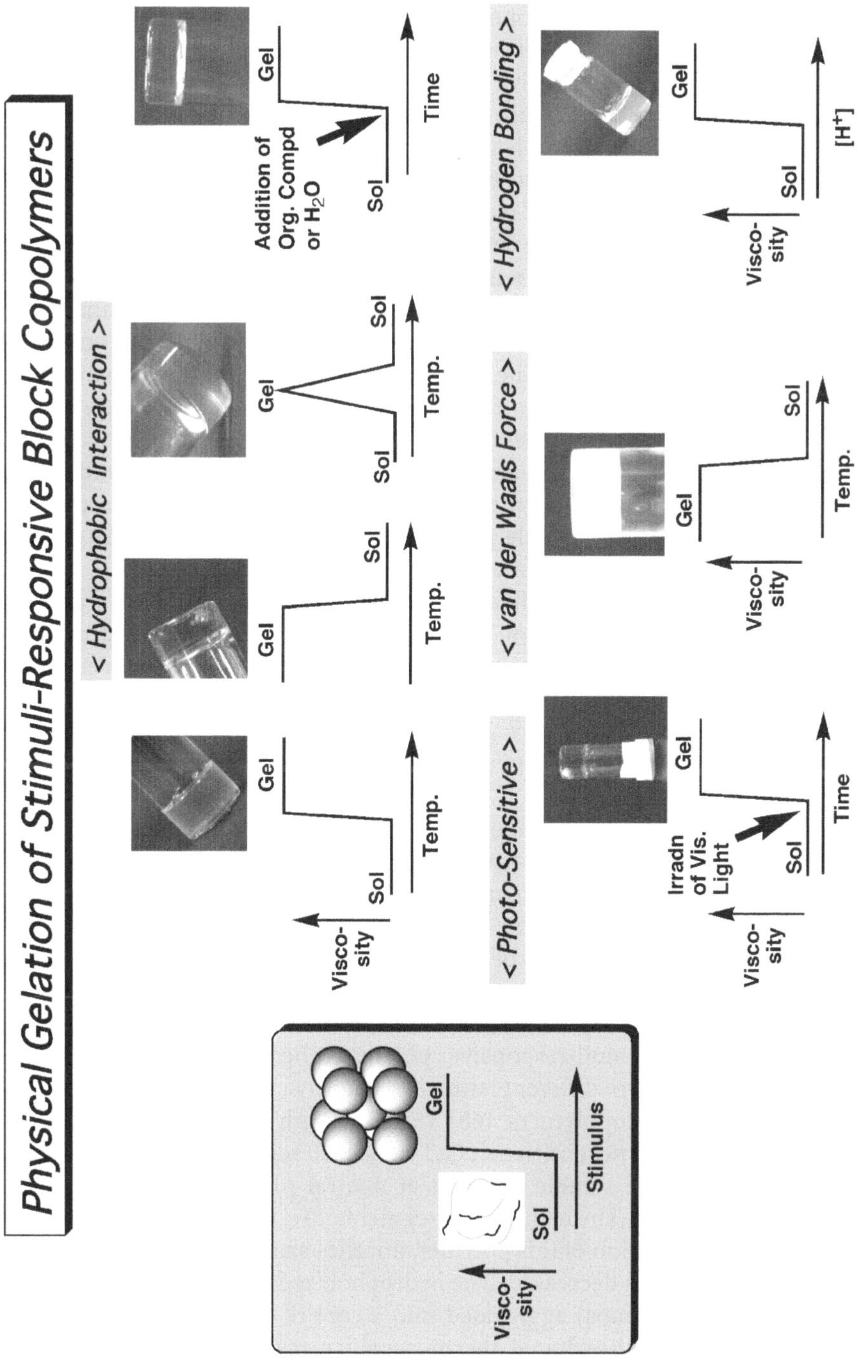

Scheme 5 Schematic illustrations of physical gelation behavior for a series of stimuli-responsive diblock copolymers and the photographs of gels

The presence of more than one stimuli-responsive site may result in unique self-association with multiple transitions. Diblock copolymers (e.g. **72**) of vinyl ethers with two thermosensitive segments possessing different critical temperatures (T_{PS}) have been synthesized by sequential living cationic polymerization [237–239]. One example is a block copolymer of 2-(2-ethoxy)ethoxyethyl vinyl ether (EOEOVE) and 2-methoxyethyl vinyl ether (MOVE) (Fig. 14), the homopolymers of which undergo phase separation in water at 41 °C and 63 °C, respectively. When an aqueous solution of a diblock copolymer (EOEOVE/MOVE = 200/400) with a narrow MWD was heated, four different viscoelastic stages were observed: clear liquid (sol, ≤ 40 °C), transparent gel (42–55 °C), hot clear liquid (sol, 57–63 °C), and opaque mixture with phase separation (> 63 °C). The temperatures of the first and third transitions corresponded to the T_{PS} values of the two segments, whereas that of the second transition was dependent on factors such as structure (sequence, composition, and MWD) and physical properties (concentration, additives). The gelation mechanism was the same as that of a single thermoresponsive diblock copolymer. The orderly packing of spherical micelles in the gel was confirmed by TEM using the freeze-fracture method [239].

Stimuli-responsive ABC triblock copolymers containing three segments with different phase separation temperatures were also synthesized [240] (Fig. 14). The triblock copolymers exhibit sensitive thermally induced physical gelation (open association) via formation of micelles. For example, an aqueous solution of **73** (m/n/l = 200/200/200) underwent multiple reversible transitions from sol (< 20 °C) to micellization (20–41 °C) to physical gelation (physical cross-linking, 41–64 °C) and finally to precipitation (> 64 °C). Furthermore, the ABC triblock copolymers exhibit Weissenberg effects in semidilute aqueous solution. In sharp contrast, another ABC triblock copolymer (**74**) with a different arrangement (m/n/l = 200/200/200), scarcely exhibits any increase in viscosity above 41 °C, although physical gelation occurred above 21 °C. The temperatures of micelle formation and physical gelation correspond to the phase-separation temperatures of the segment types in the ABC triblock copolymer.

After various stimuli-responsive polymers became available, diblock copolymers with two different stimuli-responsive segments were synthesized [225]. Block copolymers (**68**) (Fig. 12) with pH-responsive (critical pH = 6.5) and thermosensitive (LCST = 63 °C) segments ($M_n = 6.2 \times 10^4$, $M_w/M_n = 1.17$) were soluble in water at neutral pH (~ 7) and room temperature (ca. 25 °C), since the both segments are hydrophilic under these conditions. In a solution of this polymer, micelles started forming at pH 6.5 or less when the pH was decreased. The hydrophobized segments (due to undissociated carboxy groups) aggregated into a core of a micelle, with a corona consisting of the still-hydrated thermosensitive segment. Micelles with the opposite arrangement were obtained by heating the same dilute solution of **68** at pH ~ 7 above the critical temperature of poly(MOVE) (63 °C). These micellizations were confirmed by ^1H NMR in D_2O; the solubilization of a water-insoluble compound in water, monitored by UV-vis spectrometry; and dynamic light scattering (DLS) measurements.

6
New Initiating Systems and Synthetic Methodologies

The synthesis of functional or stimuli-responsive polymers in cationic polymerization often suffered due to the difficulty of polymerizing polar monomers with basic side groups. Therefore, beginning a couple of years ago, a re-examination of initiating systems was conducted. This chapter focuses on recent developments in living cationic polymerization systems, specifically the base-stabilized version. For other living cationic systems, excellent reviews are available elsewhere [215, 241–246].

6.1
"Classic" Living Cationic Polymerization with Added Base

Although esters are known to act as transfer agents in cationic polymerization, it was found that their combinations with $EtAlCl_2$ induced living cationic polymerization to yield well-defined polymers with low polydispersity [217, 247, 248]. This was the beginning of the quest for an unusual living polymerization system: a combination of a strong Lewis acid and a possible transfer agent. In a "classic" living cationic polymerization with an added base, $EtAlCl_2$ was used in nonpolar solvent such as hexane or toluene at or above 0 °C [247–249].

6.2
Recent Development of Homogeneous Catalysts

In general, in living cationic processes, stabilization of the growing active species leads to a very low concentration of real ionic species, which causes the rate of polymerization to drop. In addition, interaction between polar groups and Lewis acids retard polymerization reactions [226]. Such practical drawbacks were a bottleneck in the development of stimuli-responsive polymers. Thus, a survey of initiating systems for living cationic polymerization of vinyl ethers was carried out. According to the book "Friedel–Crafts and Related Reactions" by Olah [250], $SnCl_4$ and $FeCl_3$, compared with $AlCl_3$, have a strong affinity for Cl atoms, especially in a carbonyl-containing solvent [251].

Based on these facts, the cationic polymerization of IBVE was examined using IBVE-HCl/$SnCl_4$ (or IBEA-$EtAlCl_2$/$SnCl_4$) in toluene at 0 °C in the presence of ethyl acetate [252]. To our surprise, the polymerization rate was accelerated by a factor of 10^3, proceeding quantitatively without an induction period within only 2 min (Fig. 15) (the reaction time with $EtAlCl_2$ alone was about 1 day). Despite the rapidity of the reaction, polymerization was well controlled, giving a polymer with an extremely narrow molecular weight distribution ($M_w/M_n < 1.05$). The $SnCl_4$-induced polymerization rate for *O*- or *N*-containing monomers was shown to be 10^3-10^5 times larger than the rates using the conventional Et_xAlCl_{3-x} ($x = 1$ or 1.5) initiating systems [252]. An alternative weaker base, ethyl chloroacetate [253], which is known to induce faster reactions, realized very fast polymerization with $SnCl_4$ in toluene, being completed within 2 s (Fig. 15) (determined using a high-resolution digital video camera) [254].

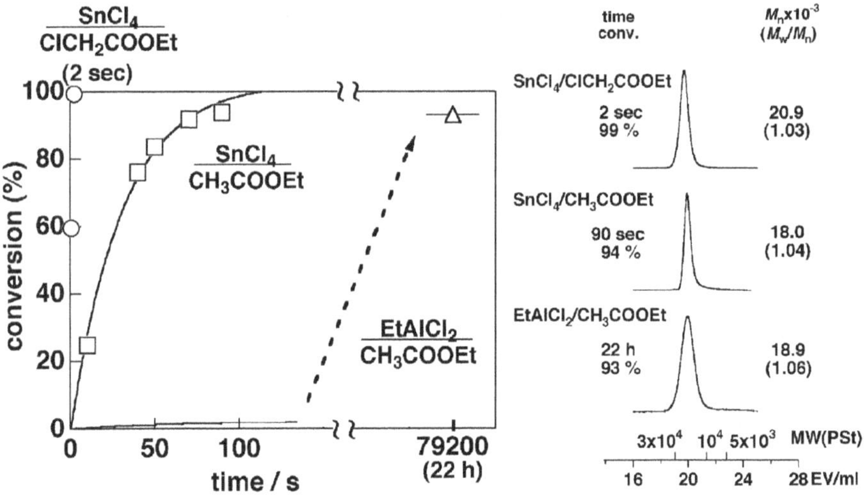

Fig. 15 Time-conversion curves and MWD profiles in the fast living cationic polymerization of IBVE

$SnCl_4$ was also used for living polymerization of various monomers containing vinyl ethers with polar side groups [252, 255, 256], α-methylvinyl ethers [257], and styrene derivatives [258], species for which it is very difficult to achieve living or even conventional cationic polymerization using an Al-based acid. In each case, the use of $SnCl_4$ significantly accelerated the reaction, inducing smooth living polymerization to give polymers with low polydispersity. The significant difference in reactivity in these cases was attributed to the hardness of the Lewis acid, $SnCl_4$ and Al-based Lewis acids.

$FeCl_3$ has advantages for industrial use: it is easy to handle [259], has low toxicity, and is economical [260]. Despite these favorable features, $FeCl_3$ has never been used for living cationic polymerization, although there have been several examples of its use in cationic polymerization of vinyl ethers [261], styrenes [262], and dienes [263]. The use of $FeCl_3$ did also result in fast living polymerization of IBVE. For example, cationic polymerization of IBVE with IBVE-HCl/$FeCl_3$ in toluene in the presence of 1,4-dioxane at 0 °C was completed in 15 s to give a polymer with a very narrow MWD (M_w/M_n = 1.06) [264]. 1,3-Dioxolane, a weaker base than 1,4-dioxane, induced faster polymerization, complete in 2–3 s.

The results obtained using $SnCl_4$ and $FeCl_3$ suggested that there could be other appropriate combinations of Lewis acids and added bases. Various metal halides comprising several main group elements and transition metals (Ti, Zr, Hf, Zn, Ga, In, Si, Ge, and Bi) were selected. The combination of each catalyst with an ester or ether resulted in living cationic polymerization [265]. All MWDs of the product polymers were very narrow, with polydispersity ranging from 1.02 to 1.10. The reaction rates differed significantly, varying from 15 s to more than 1 month. In addition, the various combinations of metal halides and bases were likely to have different functional group tolerance. These differing features offered numerous possibilities for living cationic polymerization of a new class of monomers.

6.3 Heterogeneous Living Cationic Polymerization with Fe_2O_3

During our systematic investigation of Lewis acids for cationic polymerization with added base, even some almost insoluble catalysts induced quantitative but less-controlled polymerization at a reasonable rate. Inspired by this fact, we began to examine the feasibility of living polymerization using a solid acid, which was completely insoluble in the polymerization solvent.

We first examined cationic polymerization of IBVE using iron oxide in toluene at 0 °C. In the presence of an ester or ether, the polymerization proceeded with iron oxide to over 90% conversion in 8 h to give nearly monodisperse polymers without any oligomeric byproducts. The plot of molecular weight against monomer conversion shows a linear relationship between molecular weight and conversion, which indicates that living polymerization took place [266]. To our knowledge, this is the first example of living polymerization using a heterogeneous catalyst for ionic polymerization. The catalyst was readily separated from the mixture by centrifugation, and reuse of the catalyst was examined. Up to the fifth use, the reactions gave well-defined polymers, and the product polymers were very similar in molecular weight [266].

6.4 Star-Shaped Polymers with Narrow MWDs and Gradient Copolymers

Base-stabilized living cationic polymerization permitted the design and precision synthesis of various types of polymer. The polymer linking reaction of a living polymer with a divinyl compound was known to yield star-shaped polymers with many arms [267–269]. In the reaction, the living polymers are allowed to react with a small amount of a divinyl compound to form a block copolymer, and subsequent inter- or intramolecular cross-linking reactions give a star-shaped polymer. Although this method is very effective for production of polymers with many branches and for facile arm design, all polymerization mechanisms typically resulted in relatively low yields and broad MWDs [267–275].

The base-stabilizing process has greater stability of the growing ends, relative to other living cationic systems, such as the "counter-ion" stabilizing systems. Therefore, we examined star polymer synthesis using various strong Lewis acids such as organoaluminum halide ($EtAlCl_2$) or titanium tetrachloride ($TiCl_4$) in the presence of a weak Lewis base. A Lewis base not only stabilizes carbocations, but also promotes assembly of the propagation species as an intermolecular linking reaction, resulting in the construction of a micelle-like association of living polymers in the reaction mixture.

Star-shaped poly(IBVE) was obtained in 100% yield [276] from the living polymer reaction ($DP_n = 50–300$) of IBVE, prepared with IBEA/$EtAlCl_2$ at 0 °C in hexane in the presence of ethyl acetate, with a small amount of 1,4-cyclohexanedimethanol divinyl ether. A notable feature of these star-shaped polymers was their extremely narrow MWDs ($M_w/M_n = 1.1–1.2$) [276]. To the best of our knowledge, this was the first example of selective preparation of star-shaped polymers with a narrow MWD in quantitative yield, which had never before been achieved even with other polymerization mechanisms. The M_w ranged from 6×10^4 to 30×10^4, and each polymer molecule had between 9 and 44 arms.

Thermoresponsive star-shaped polymers were also prepared in quantitative yield [276, 277]. Star block copolymers also induced reversible physical gelation at a higher concentration. A 10% aqueous solution of the star block copolymer with EOVE and MOVE segments in the outer and inner layers, respectively, underwent rapid physical gelation upon heating. Intermolecular aggregation of the outer segments was induced upon heating, resulting in the formation of three-dimensional physical network. The star block copolymer with the opposite arrangement underwent sol–gel transition upon cooling in the 15 wt % aqueous solution. This transition is presumably attributed to the change in diameter of star molecules, driven by the hydration or dehydration of poly(EOVE) segments. A similar sol–gel phenomenon was observed with linear double thermoresponsive diblock copolymers, which form spherical micelles with the EOVE segments in the inner layer [239].

For stimuli-responsive self-assembly systems using thermoresponsive polymers, precision control of the primary polymer structure was essential in order to achieve high sensitivity in the phase transition and unique self-assembly behavior. Most of the examples shown above for self-assembly have been block copolymers with narrow MWDs. These polymers were prepared by a "static" living polymerization method (batch method), as is usually used in living systems, but flow reaction systems can also be used for living polymerization. For example, various sequence-controlled polymers may be prepared by gradually feeding a different monomer into a polymerization mixture (gradient copolymer), or successively adding the mixture to a terminating agent or another monomer (MWD and/or segment length distribution-controlled polymer).

Synthesis of a gradient copolymer was carried out using $IBEA/Et_{1.5}AlCl_{1.5}$ in the presence of ethyl acetate in toluene at 0 °C [278]. Shortly after initiation of living polymerization of the first monomer, EOVE, a second monomer, MOVE, was added continuously with a syringe pump until the conversion of EOVE reached about 90%. Polymerization proceeded in a living fashion to give a polymer of narrow MWD with a compositional change along the polymer chain ($M_n = 4.9 \times 10^4$, $M_w/M_n = 1.25$). The composition ratio and the instantaneous composition, determined by ^{1}H NMR spectroscopy and gravimetry, agreed with the theoretical values.

Dynamic light scattering and small-angle neutron scattering investigations were carried out for aqueous solutions of the gradient copolymer consisting of EOVE and MOVE [278]. Although aqueous solutions of EOVE homopolymer precipitated at 20 °C, the copolymer system showed a different temperature dependence. A 0.3 wt % gradient system had a micellization temperature range between 20 and 30 °C. The corresponding block copolymer system underwent a sharp transition at around 20 °C, as observed for the EOVE homopolymer. Micellization was continuous in the gradient system, whereas a stepwise transition was observed in the block system. The size of the resulting micelles in the gradient system was smaller than that in the block system. The continuous change and smaller micelle size observed in the gradient copolymer system were explained by a "reel-in" effect in the micellization process [278].

MWD-controlled diblock copolymers were prepared by gradually feeding a polymerization mixture into a sample of a terminating agent during second-stage polymerization [279]. The resulting polymers contain a nearly monodisperse segment and a segment with a preset length distribution. Micellization of these polymers in water demonstrated that the length distribution of the core-forming segment is a decisive factor in determining the size and/or size distribution of the micelles. The uniform length of the core segment resulted in the formation of narrowly distributed micelles, despite differing corona segment lengths. Furthermore, a diblock copolymer with narrow MWD but various compositions was prepared by gradually feeding a polymerization mixture into another monomer solution as first-stage polymerization proceeded [279]. Such diblock copolymers with various compositions showed different micellization behaviors, indicative of the formation of novel assembling structures.

7
Conclusion

The synthesis based on living polymerization allowed the preparation of various designed stimuli-responsive polymers, which can hardly be synthesized without living polymerization. As demonstrated in this review, studies on the precision synthesis of various NIPAM-based polymers have been increasing exponentially since the achievement of living radical polymerization of NIPAM. This recent great advance in the synthesis has made a tremendous impact on the fields relating to stimuli-responsive materials science. The successful synthesis of various stimuli-responsive poly(vinyl ether)s including block and star-shaped polymers via base-stabilized living cationic polymerization is also described. The studies on the poly(vinyl ether) systems demonstrated that judicious design of polymer structures realized reversible and sensitive self-association of block and star polymers with narrow MWD and controlled sequences. The results with poly(vinyl ether)s provide a guidance framework for future research on the construction of more sophisticated self-assembling systems based on stimuli-responsive polymers.

Acknowledgements We thank all the students in our group and former members who have contributed to the developments of stimuli-responsive polymers and living polymerization systems summarized in this review, especially Shinji Sugihara, Tomohide Yoshida, and Takaho Shibata. Yukari Oda and Hiroaki Shimomoto are particularly appreciated for help with the preparation of this review. We would also like to acknowledge Professor Shibayama and his co-workers for collaborative scattering experiments with the physical gels obtained from block copolymers.

References

1. Hoffman AS (1995) Macromol Symp 98:645
2. Hoffman AS, Stayton PS (2004) Macromol Symp 207:139
3. Kikuchi A, Okano T (2002) Prog Polym Sci 27:1165
4. Nath N, Chilkoti A (2002) Adv Mater 14:1243
5. Ito T, Hioki T, Yamaguchi T, Shinbo T, Nakao S, Kimura S (2002) J Am Chem Soc 124:7840
6. Tsutsui H, Akashi R (2006) J Appl Polym Sci 102:362
7. Mori T, Maeda M (2002) Polym J 34:624
8. Heymann E (1935) Trans Faraday Soc 31:846
9. Chakhovsky N, Martin RH, Neckel R (1956) Bull Soc Chim Belges 65:453
10. Horne RA, Almeida JP, Day AF, Yu N-T (1971) Colloid Interface Sci 35:77
11. Nord FF, Bier M, Timasheff SN (1951) J Am Chem Soc 73:289
12. Scarpa JS, Mueller DD, Klotz IM (1967) J Am Chem Soc 89:6024
13. Ito S (1989) Kobunshi Ronbunshu 46:437
14. Schild HG (1992) Prog Polym Sci 17:163
15. Heskins M, Guilet J (1968) J Macromol Sci Chem A2:1441
16. Fujishige S, Kubota K, Ando I (1989) J Phys Chem 93:3311
17. Jeong B, Gutowska A (2002) Trend Biotechnol 20:305

18. Gil ES, Hudson SM (2004) Prog Polym Sci 29:1173
19. Matyjaszewski K (1998) ACS Symp Ser 685, Controlled radical polymerization. ACS, Washington, DC
20. Matyjaszewski K (2000) ACS Symp Ser 768, Controlled/living radical polymerization. ACS, Washington, DC
21. Matyjaszewski K (2003) ACS Symp Ser 854, Advances in controlled/living radical polymerization. ACS, Washington, DC
22. McCormik CL (2001) ACS Symp Ser 780, Stimuli-responsive water soluble and amphiphilc polymers. ACS, Washington, DC
23. Lazzari M, Lin G, Lecommandoux S (2006) Block copolymers in nanoscience. Wiley-VCH, Weinheim
24. Kawaguchi H (2000) Prog Polym Sci 25:1171
25. Nath N, Chilkoti A (2002) Adv Mater 14:1243
26. Matyjaszewski K (2005) Prog Polym Sci 30:858
27. Otsu T, Yoshida M (1982) Makromol Chem Rapid Commun 3:127
28. Georges MK, Veregin RPN, Kazmaier PM, Hamer GK (1993) Macromolecules 26:2987
29. Hawker CJ (1994) J Am Chem Soc 116:11185
30. Studer A, Schulte T (2005) Chem Record 5:27
31. Kato M, Kamigaito M, Sawamoto M, Higashimura T (1995) Macromolecules 28:1721
32. Wang JS, Matyjaszewski K (1995) J Am Chem Soc 117:5614
33. Percec V, Guliashvili T, Ladislaw JS, Wistrand A, Stjerndahl A, Sienkowska MJ, Monteiro MJ, Sahoo S (2006) J Am Chem Soc 128:14156
34. Chiefari J, Chong YK, Ercole F, Krstina J, Jeffery J, Le TPT, Mayadunne RTA, Meijs GF, Moad CL, Moad G, Rizzardo E, Thang SH (1998) Macromolecules 31:5559
35. Moad G, Rizzardo E, Thang SH (2005) Aust J Chem 58:379
36. Perrier S, Takolpuckdee P (2005) J Polym Sci Part A Polym Chem 43:5347
37. Barner-Kowollik C, Davis TP, Heuts JPA, Stenzel MH, Vana P, Whittaker M (2003) J Polym Sci Part A Polym Chem 41:365
38. Savariar EN, Thayumanavan S (2004) J Polym Sci Part A Polym Chem 42:6340
39. Ganachaud F, Monteiro MJ, Gilbert RG, Dourges M-A, Thang SH, Rizzardo E (2000) Macromolecules 33:6738
40. Miwa N, Fukuda T, Minoda M (2000) Polym Prepr Jpn 49:1737
41. Schilli C, Lanzendörfer MG, Müller AHE (2002) Macromolecules 35:6819
42. Convertine AJ, Ayres N, Scales CW, Lowe AB, McCormick CL (2004) Biomacromolecules 5:1177
43. Goto A, Kwak Y, Fukuda T, Yamago S, Iida K, Nakajima M, Yoshida J (2003) J Am Chem Soc 125:8720
44. Yamago S, Ray B, Iida K, Yoshida J, Tada T, Yoshizawa K, Kwak Y, Goto A, Fukuda T (2004) J Am Chem Soc 126:13908
45. Yamago S (2006) J Polym Sci Part A Polym Chem 44:1
46. Millard P-E, Barner L, Stenzel MH, Davis TP, Barner-Kowollik C, Müller AHE (2006) Macromol Rapid Commun 27:821
47. LeMieux MC, Peleshanko S, Lin Y-H, Tsukruk VV (2005) Polym Prepr 46(2):20
48. Ishizu K, Khan RA, Furukawa T, Furo M (2004) J Appl Polym Sci 91:3233
49. Ray B, Isobe Y, Morioka K, Habaue S, Okamoto Y, Kamigaito M, Sawamoto M (2003) Macromolecules 36:543
50. Ray B, Isobe Y, Morioka K, Habaue S, Okamoto Y, Kamigaito M, Sawamoto M (2004) Macromolecules 37:1702

51. Ray B, Okamoto Y, Kamigaito M, Sawamoto M, Seno K, Kanaoka S, Aoshima S (2005) Polym J 37:234
52. Sugiyama Y, Satoh K, Kamigaito M, Okamoto Y (2006) J Polym Sci Part A Polym Chem 44:2086
53. Harth E, Bosman AW, Benoit D, Helms B, Fréchet JMJ, Hawker CJ (2001) Macromol Symp 174:85
54. Bosman AW, Vestberg R, Heumann A, Fréchet JMJ, Hawker CJ (2003) J Am Chem Soc 125:715
55. Schulte T, Siegenthaler KO, Luftmann H, Letzel M, Studer A (2005) Macromolecules 38:6833
56. Masci G, Giacomelli L, Crescenzi V (2004) Macromol Rapid Commun 25:559
57. Xia Y, Yin X, Burke NAD, Stöver HDH (2005) Macromolecules 38:5937
58. Yohannes G, Shan J, Jussila M, Nuopponen M, Tenhu H, Riekkola M-L (2005) J Sep Sci 28:435
59. Schilli CM, Müller AHE, Rizzardo E, Thang SH, Chong YK (2003) In: Matyjaszewski K (ed) ACS Symp Ser 854, Advances in controlled/living radical polymerization. ACS, Washington, DC, p 603
60. Ishizone T, Ito M (2002) J Polym Sci Part A Polym Chem 40:4328
61. Ito M, Ishizone T (2006) J Polym Sci Part A Polym Chem 44:4832
62. Kitayama T, Shibuya W, Katsukawa K (2002) Polym J 34:405
63. Kobayashi M, Ishizone T, Nakahama S (2000) J Polym Sci Part A Polym Chem 38:4677
64. Topp MDC, Dijkstra PJ, Talsma H, Feijen J (1997) Macromolecules 30:8518
65. Lin H-H, Cheng Y-L (2001) Macromolecules 34:3710
66. Yu H, Feng Z-G, Zhang A-Y (2006) J Polym Sci Part A Polym Chem 44:3717
67. Motokawa R, Morishita K, Koizumi S, Nakahira T, Annaka M (2005) Macromolecules 38:5748
68. Cammas S, Suzuki K, Sone C, Sakurai Y, Kataoka K, Okano T (1997) J Controlled Release 48:157
69. Kohori F, Sakai K, Aoyagi T, Yokoyama M, Sakurai Y, Okano T (1998) J Controlled Release 55:87
70. Yoshida R, Uchida K, Kaneko Y, Sakai K, Kikuchi A, Sakurai Y, Okano T (1995) Nature 374:240
71. Gohy J-F (2005) Adv Polym Sci 190:65
72. Liu B, Perrier S (2005) J Polym Sci Part A Polym Chem 43:3643
73. Convertine AJ, Lokitz BS, Vasileva Y, Myrick LJ, Scales CW, Lowe AB, McCormick CL (2006) Macromolecules 39:1724
74. Licea-Claverie A, Carrión-Garcia SA, Medina-Urquiza MR, Cornejo-Bravo JM, Hawker CJ, Frank CW (2006) Polym Mat: Sci Eng 51:170
75. Skrabania K, Kristen J, Laschewsky A, Akdemir Ö, Hoth A, Lutz J-F (2007) Langmuir 23:84
76. Mertoglu M, Garnier S, Laschewsky A, Skrabania K, Strosberg J (2005) Polymer 46:7726
77. Cao Y, Zhu XX (2006) Polym Prepr 47(1):354
78. Schilli CM, Zhang M, Rizzardo E, Thang SH, Chong YK, Edwards K, Karlsson G, Müller AHE (2004) Macromolecules 37:7861
79. Yang C, Cheng Y-L (2006) J Appl Polym Sci 102:1191
80. Li G, Shi L, An Y, Zhang W, Ma R (2006) Polymer 47:4581
81. Yusa S, Shimada Y, Mitsukami Y, Yamamoto T, Morishima Y (2004) Macromolecules 37:7507

82. Masci G, Giacomelli L, Crescenzi V (2005) J Polym Sci Part A Polym Chem 43:4446
83. Zhang P, Liu Q, Qing A, Shi J, Lu M (2006) J Polym Sci Part A Polym Chem 44:3312
84. Xu F-J, Kang E-T, Neoh E-T (2006) Biomaterials 27:2787
85. Nuopponen M, Ojala J, Tenhu H (2004) Polymer 45:3643
86. Gibbons O, Carroll WM, Aldabbagh F, Yamada B (2006) J Polym Sci Part A Polym Chem 44:6410
87. Tang T, Castelletto V, Parras P, Hamley IW, King SM, Roy D, Perrier S, Hoogenboom R, Schubert US (2006) Macromol Chem Phys 207:1718
88. Hong C-Y, You Y-Z, Pan C-Y (2004) J Polym Sci Part A Polym Chem 42:4873
89. Zhang W, Shi L, Wu K, An Y (2005) Macromolecules 38:5743
90. Qin S, Geng Y, Discher DE, Yang S (2005) Polym Mat: Sci Eng 93:191
91. Virtanen J, Holappa S, Lemmetyinen H, Tenhu H (2002) Macromolecules 35:4763
92. You Y, Hong C, Wang W, Lu W, Pan C-Y (2004) Macromolecules 37:9761
93. Hales M, Barner-Kowollik C, Davis TP, Stenzel MH (2004) Langmuir 20:10809
94. Li C, Tang Y, Armes SP, Morris CJ, Rose SF, Lloyd AW, Lewis AL (2005) Biomacromolecules 6:994
95. Kulkarni S, Schilli C, Müller AHE, Hoffman AS, Stayton PS (2004) Bioconjugate Chem 15:747
96. Kulkarni S, Schilli C, Grin B, Müller AHE, Hoffman AS, Stayton PS (2006) Biomacromolecules 7:2736
97. Heredia KL, Bontempo D, Ly T, Byers JT, Halstenberg S, Maynard HD (2005) J Am Chem Soc 127:16955
98. Bontempo D, Li RC, Ly T, Brubaker CE, Maynard HD (2005) Chem Commun, p 4702
99. Bontempo D, Maynard HD (2005) J Am Chem Soc 127:6508
100. Bontempo D, Maynard HD (2005) Polym Prepr 46(1):78
101. Bernard J, Hao X, Davis TP, Barner-Kowollik C, Stenzel MH (2006) Biomacromolecules 7:232
102. Stenzel MH, Zhang L, Huck WTS (2006) Macromol Rapid Commun 27:1121
103. Kuroda K, Swager TM (2004) Macromolecules 37:716
104. Kuroda K, Swager TM (2003) Chem Comm, p 26
105. Zhang Y, Luo S, Liu S (2005) Macromolecules 38:9813
106. Virtanen J, Arotçaréna M, Heise B, Ishaya S, Laschewsky A, Tenhu H (2002) Langmuir 18:5360
107. Arotçaréna M, Heise B, Ishaya S, Laschewsky A (2002) J Am Chem Soc 124:3787
108. Han D-H, Pan C-Y (2006) Macromol Chem Phys 207:836
109. Lokitz BS, Convertine AJ, Ezell RG, Heidenreich A, Li Y, McCormick CL (2006) Macromolecules 39:8594
110. Li C, Buurma NJ, Haq I, Turner C, Armes SP, Castelletto V, Hamley IW, Lewis AL (2005) Langmuir 21:11026
111. Li Y, Lokitz BS, McCormick CL (2006) Macromolecules 39:81
112. Ge Z, Luo S, Liu S (2006) J Polym Sci Part A Polym Chem 44:1357
113. Moad G, Chong YK, Postma A, Rizzardo E, Thang SH (2005) Polymer 46:8458
114. Nakayama M, Okano T (2005) Biomacromolecules 6:2320
115. Xia Y, Burke NAD, Stöver HDH (2006) Macromolecules 39:2275
116. Postma A, Davis TP, Li G, Moad G, O'Shea MS (2006) Macromolecules 39:5307
117. Scales CW, Convertine AJ, McCormick CL (2006) Polym Prepr 46(2):393
118. Scales CW, Convertine AJ, McCormick CL (2006) Biomacromolecules 7:1389
119. Hong C-Y, Pan C-Y (2006) Macromolecules 39:3517
120. Kujawa P, Segui F, Shaban S, Diab C, Okada Y, Tanaka F, Winnik FM (2006) Macromolecules 39:341

121. Kujawa P, Watanabe H, Tanaka F, Winnik FM (2005) Eur Phys J E17:129
122. Qiu X-P, Winnik FM (2006) Macromol Rapid Commun 27:1648
123. Duan Q, Narumi A, Miura Y, Shen X, Sato S, Satoh T, Kakuchi T (2006) Polymer J 38:306
124. Zhou G, Harruna II, Ingram CW (2005) Polymer 46:10672
125. Siegwart DJ, Hollinger JO, Matyjaszewski K (2006) Polym Mat: Sci Eng 51:823
126. Yin X, Hoffman AS, Stayton PS (2006) Biomacromolecules 7:1381
127. Metz N, Theato P (2006) Polym Prepr 47(2):716
128. You Y-Z, Hong C-Y, Pan C-Y, Wang P-H (2004) Adv Mater 16:1953
129. Zheng Q, Pan C-Y (2006) Eur Polym J 42:807
130. Luo S, Xu J, Zhu Z, Wu C, Liu S (2006) J Phys Chem B 110:9132
131. Xu J, Luo S, Shi W, Liu S (2006) Langmuir 22:989
132. Carter S, Hunt B, Rimmer S (2005) Macromolecules 38:4595
133. Haba Y, Harada A, Takagishi T, Kono K (2004) J Am Chem Soc 126:12760
134. Zheng G, Pan C-Y (2005) Polymer 46:2802
135. Lambeth RH, Ramakrishnan S, Mueller R, Poziemski JP, Miguel GS, Markoski LJ, Zukoski CF, Moore JS (2006) Langmuir 22:6352
136. Chen M, Ghiggino KP, Thang SH, Wilson GJ (2005) Polym Int 55:757
137. Chen M, Ghiggino KP, Thang SH, Wilson GJ (2005) Angew Chem Int Ed 44:4368
138. Yang L-P, Pan C-Y (2006) Aust J Chem 59:733
139. Plummer R, Hill DJT, Whittaker AK (2006) Macromolecules 39:8379
140. Wan D, Fu Q, Huang J (2005) J Polym Sci Part A Polym Chem 43:5652
141. Carter SR, Rimmer S, Rutkaite R, Swanson L, Fairclough JPA, Sturdy A, Webb M (2006) Biomacromolecules 7:1124
142. Carter SR, Hunt B, Rimmer S, Webb M, Sturdy A (2005) Polym Prepr 46(2):253
143. Carter SR, Rimmer S, Rutkaite R, Swanson L, Haycock J (2005) Polym Mat: Sci Eng 93:503
144. Carter SR, Hunt B, Rimmer S (2005) Macromolecules 38:4595
145. Feng X-S, Pan C-Y (2002) Macromolecules 35:4888
146. Hu T, You Y, Pan C-Y, Wu C (2002) J Phys Chem B 106:6659
147. Cui Y, Tao C, Zheng S, He Q, Ai S, Li J (2005) Macromol Rapid Commun 26:1552
148. Tsuji S, Kawaguchi H (2004) Langmuir 20:2449
149. Tsuji S, Kawaguchi H (2005) Langmuir 21:2434
150. Suzuki D, Tsuji S, Kawaguchi H (2005) Chem Lett 34:242
151. Lu Y, Wittemann A, Ballauff M, Drechsler M (2006) Macromol Rapid Commun 27:1137
152. Bontempo D, Masci D, De Leonardis P, Mannina L, Capitani D, Crescenzi V (2006) Biomacromolecules 7:2154
153. Kim DJ, Heo JY, Kim KS, Choi IS (2003) Macromol Rapid Commun 24:517
154. Couet J, Biesalski M (2006) Macromolecules 39:7258
155. Kizhakkedathu JN, Norris-Jones R, Brooks DE (2004) Macromolecules 37:734
156. Farhan T, Huck WTS (2004) Eur Polym J 40:1599
157. Desai SM, Solanky SS, Mandale AB, Rathore K, Singh RP (2003) Polymer 44:7645
158. Ying L, Yu WH, Kang ET, Neoh KG (2004) Langmuir 20:6032
159. Li C, Gunari N, Fischer K, Janshoff A, Schmidt M (2004) Angew Chem Int Ed 43: 1101
160. Liu Q, Zhang P, Qing A, Lan Y, Lu M (2006) Polymer 47:2330
161. Liu Q, Zhang P, Qing A, Lan Y, Shi J, Lu M (2006) Polymer 47:6963
162. Liu Q, Zhang P, Lu M (2005) J Polym Sci Part A Polym Chem 43:2615

163. Shan J, Nuopponen M, Jiang H, Kauppinen E, Tenhu H (2003) Macromolecules 36:4526
164. Kaholek M, Lee W-K, Ahn S-J, Ma H, Caster KC, LaMattina B, Zauscher S (2004) Chem Mater 16:3688
165. Wang X, Tu H, Braun PV, Bohn PW (2006) Langmuir 22:817
166. Lokuge I, Wang X, Bohn PW (2007) Langmuir 23:305
167. Zhang Y, Qin S, Taylor JA, Aizenberg J, Yang S (2006) Polym Mat: Sci Eng 94:852
168. Zhou F, Huck WTS (2005) Chem Commun, p 5999
169. Yim H, Kent MS, Mendez S, Balamurugan SS, Balamurugan S, Lopez GP, Satija S (2004) Macromolecules 37:1994
170. Kaholek M, Lee W-K, LaMattina B, Caster KC, Zauscher S (2004) Nano Lett 4:373
171. Yim H, Kent MS, Mendez S, Lopez GP, Satija S, Seo Y (2006) Macromolecules 39:3420
172. Teare DOH, Barwick DC, Schofield WCE, Garrod RP, Ward LJ, Badyal JPS (2005) Langmuir 21:11425
173. Zhu M-Q, Wang L-Q, Exarhos GJ, Li ADQ (2004) J Am Chem Soc 126:2656
174. Kim DJ, Kang SM, Kong B, Kim W-J, Paik H-J, Choi H, Choi IS (2005) Macromol Chem Phys 206:1941
175. Shan J, Nuopponen M, Jiang H, Kauppinen E, Tenhu H (2003) Macromolecules 36:4526
176. Seino M, Yokomachi K, Hayakawa T, Kikuchi R, Kakimoto M, Horiuchi S (2006) Polymer 47:1946
177. Tu H, Heitzman CE, Braun PV (2004) Langmuir 20:8313
178. Xu FJ, Zhong SP, Yung LYL, Tong YW, Kang E-T, Neoh KG (2006) Biomaterials 27:1236
179. Xu FJ, Zhong SP, Yung LYL, Kang ET, Neoh KG (2004) Biomacromolecules 5:2392
180. Li J, Chen X, Chang Y-C (2005) Langmuir 21:9562
181. Fu Q, Rao GVR, Basame SB, Keller DJ, Artyushkova K, Fulghum JE, Lopez GP (2004) J Am Chem Soc 126:8904
182. Idota N, Kikuchi A, Kobayashi J, Akiyama Y, Sakai K, Okano T (2006) Langmuir 22:425
183. Liu Y, Klep V, Luzinov I (2006) J Am Chem Soc 128:8106
184. Hong C-Y, You Y-Z, Pan C-Y (2005) Chem Mater 17:2247
185. Sun T, Liu H, Song W, Wang X, Jiang L, Li L, Zhu D (2004) Angew Chem Int Ed 43:4663
186. Kong H, Li W, Gao C, Yan D, Jin Y, Walton DRM, Kroto HW (2004) Macromolecules 37:6683
187. Xu G, Wu W-T, Wang Y, Pang W, Wang P, Zhu Q, Lu F (2006) Nanotechnol 17:2458
188. Tsubokawa N (2002) Bull Chem Soc Jpn 75:2115
189. Haraguchi K, Takehisa T, Fan S (2002) Macromolecules 35:10162
190. Nijenhuis K (1997) Adv Polym Sci 130:1
191. Hamely IW (2005) Block copolymers in solution. Wiley, Chichester
192. Wanka G, Hoffman H, Ulbricht W (1994) Macromolecules 27:4145
193. Svensson M, Alexandridis P, Linse P (1999) Macromolecules 32:637
194. Mortensen K, Pedersen JS (1993) Macromolecules 26:805
195. Jeong B, Bae YH, Lee DS, Kim SW (1997) Nature 388:860
196. Li H, Yu G-E, Price C, Booth C, Hecht E, Hoffmann H (1997) Macromolecules 30:1347
197. Mortensen K (1993) Prog Colloid Polym Sci 93:72
198. Mortensen K, Brown W, Jørgensen E (1994) Macromolecules 27:5654

199. Alexandridis P, Hatton TA (1995) Colloid Surface A: Physicochem Eng Asp 96:1
200. Mortensen K (2001) Colloid Surface A: Physicochem Eng Asp 183–185:277
201. Han S, Hagiwara M, Ishizone T (2003) Macromolecules 36:8312
202. Aoki S, Koide A, Imabayashi S, Watanabe M (2002) Chem Lett, p 1128
203. Lee BH, Lee YM, Sohn YS, Song S-C (2002) Macromolecules 35:3876
204. Uyama H, Kobayashi S (1992) Chem Lett 21:1643
205. Akashi M, Nakano S, Kishida A (1996) J Polym Sci Part A Polym Chem 34:301
206. Aoki T, Muramatsu M, Torii T, Sanui K, Ogata N (2001) Macromolecules 34:3118
207. Ebara M, Aoyagi T, Sakai K, Okano T (2000) Macromolecules 33:8312
208. Nichifor M, Zhu XX (2003) Polymer 44:3053
209. Furukawa H, Onishi N, Kataoka K, Ueno K (2000) Polym Prepr Jpn 49:3079
210. Onishi N, Furukawa H, Kondo A (2003) Ohyoubutsuri 72:909
211. Buscall R, Corner T (1982) Eur Polym J 18:967
212. Klenina OV, Fain EG (1981) Polym Sci USSR 23:1439
213. Aoki T, Nakamura K, Sanui K, Kikuchi A, Okano T, Sakurai Y, Ogata N (1999) Polym J 31:1185
214. Schulz DN, Petiffer DG, Agarwal PK, Larabee J, Kaladas JJ, Soni L, Handweker B, Garner RT (1986) Polymer 27:1734
215. Matyjaszewski K (ed) (1996) Cationic polymerizations: mechanisms, synthesis, and applications, Chap 4 and 5. Marcel Dekker, New York
216. Aoshima S, Sugihara S, Shibayama M, Kanaoka S (2004) Macromol Symp 215:151
217. Aoshima S, Yoshida T, Kanazawa A, Kanaoka S (2007) J Polym Sci Part A Polym Chem (Highlight) 45:1801
218. Aoshima S, Oda H, Kobayashi E (1992) J Polym Sci Part A Polym Chem 30:2407
219. Aoshima S, Oda H, Kobayashi E (1992) Kobunshi Ronbunshu 49:933
220. Sugihara S, Hashimoto K, Matsumoto Y, Kanaoka S, Aoshima S (2003) J Polym Sci Part A Polym Chem 41:3300
221. Sugihara S, Kanaoka S, Aoshima S (2004) Macromolecules 37:1711
222. Seno K, Inaoka M, Kanaoka S, Aoshima S (2004) Polym Prepr 45:632
223. Yoshida T, Seno K, Kanaoka S, Aoshima S (2005) J Polym Sci Part A Polym Chem 43:1155
224. Date A, Kanaoka S, Kato T, Aoshima S (2005) Polym Prepr 46:977
225. Tsujino T, Kanaoka S, Aoshima S (2005) Polym Prepr 46:865
226. Yoshida T, Kanaoka S, Aoshima S (2005) J Polym Sci Part A Polym Chem 43:5138
227. Yoshida T, Kanaoka S, Aoshima S (2005) J Polym Sci Part A Polym Chem 43:4292
228. Yoshida T, Kanaoka S, Aoshima S (2005) J Polym Sci Part A Polym Chem 43:5337
229. Sugihara S, Matsuzono S, Sakai H, Abe M, Aoshima S (2001) J Polym Sci Part A Polym Chem 39:3190
230. Fuse C, Okabe S, Shibayama M, Sugihara S, Aoshima S (2004) Macromolecules 37:7791
231. Tsubouchi S, Kanaoka S, Aoshima S (2003) Polym Prepr Jpn 52:1325
232. Osaka N, Okabe S, Karino T, Shibayama M, Hirabaru Y, Aoshima S (2006) Macromolecules 39:5875
233. Hashimoto K, Aoshima S (2001) J Polym Sci Part A Polym Chem 39:746
234. Sugihara S, Hashimoto K, Okabe S, Shibayama M, Kanaoka S, Aoshima S (2004) Macromolecules 37:336
235. Okabe S, Sugihara S, Aoshima S, Shibayama M (2002) Macromolecules 35:8139
236. Okabe S, Sugihara S, Aoshima S, Shibayama M (2003) Macromolecules 36:4099
237. Aoshima S, Sugihara S (2001) Kobunshi Ronbunshu 58:304
238. Aoshima S, Sugihara S (2000) J Polym Sci Part A Polym Chem 38:3962

239. Sugihara S, Kanaoka S, Aoshima S (2005) Macromolecules 38:1919
240. Sugihara S, Kanaoka S, Aoshima S (2004) J Polym Sci Part A Polym Chem 42:2601
241. Kennedy JP, Iván B (1992) Designed polymers by carbocationic macromolecular engineering: theory and practice. Hanser Publishers, München
242. Kennedy JP (1999) J Polym Sci Part A Polym Chem 37:2285
243. Kennedy JP (2005) J Polym Sci Part A Polym Chem 43:2951
244. Puskas JE, Chen Y, Dahman Y, Padavan D (2004) Biomacromolecules 5:1142
245. Puskas JE, Chen Y (2004) J Polym Sci Part A Polym Chem 42:3091
246. Sawamoto M (1991) Prog Polym Sci 16:111
247. Aoshima S, Higashimura T (1986) Polym Bull 15:417
248. Aoshima S, Higashimura T (1989) Macromolecules 22:1009
249. Aoshima S, Kishimoto Y, Higashimura T (1989) Macromolecules 22:3877
250. Olah GA (ed) (1963) Friedel–Crafts and related reactions. Interscience, New York
251. Gutmann V, Hampel G (1961) Monatsh Chem 92:1048
252. Yoshida T, Tsujino T, Kanaoka S, Aoshima S (2005) J Polym Sci Part A Polym Chem 43:468
253. Aoshima S, Shachi Y, Kobayashi E (1991) Makromol Chem 192:1759
254. Yoshida T, Kanazawa A, Kanaoka S, Aoshima S (2005) J Polym Sci Part A Polym Chem 43:4288
255. Oda Y, Tsujino T, Yonezumi M, Sasai A, Kanaoka S, Aoshima S (2006) Polym Prepr Jpn 55:2533
256. Yonezumi M, Takaku R, Kanaoka S, Aoshima S (2006) Polym Prepr Jpn 55:2521
257. Takaku R, Kanaoka S, Aoshima S (2006) Proceedings of International Symposium on Advanced Polymers for Emerging Technology 615
258. Yamamoto H, Kanaoka S, Aoshima S (2006) Polym Prepr Jpn 55:2801
259. Aoyama N, Manabe K, Kobayashi S (2004) Chem Lett 33:312
260. Bolm C, Legros J, Paih JL, Zani L (2004) Chem Rev 104:6217
261. Matsuzaki K, Hamada M, Arita K (1967) J Polym Sci Part A-1 5:1233
262. Sakurada Y, Higashimura T, Okamura S (1958) J Polym Sci 33:496
263. Santarella JM, Rousset E, Randriamahefa S, Macedo A, Cheradame H (2000) Eur Polym J 36:2715
264. Kanazawa A, Hirabaru Y, Kanaoka S, Aoshima S (2006) J Polym Sci Part A Polym Chem 44:5795
265. Kanazawa A, Kanaoka S, Aoshima S (2006) Polym Prepr 47:135
266. Kanazawa A, Kanaoka S, Aoshima S (2007) J Am Chem Soc 129:2420
267. Bauer BJ, Fetters LJ (1978) Rubber Chem Technol 51:406
268. Bywater S (1979) Adv Polym Sci 30:89
269. Hadjichristidis N (1999) J Polym Sci Part A Polym Chem 37:857
270. Zhang X, Xia J, Matyjazewski K (2000) Macromolecules 33:2340
271. Baek KY, Kamigaito M, Sawamoto M (2001) Macromolecules 34:215
272. Bosman AW, Heumann A, Klaerner G, Fréchet JMJ, Hawker CJ (2001) J Am Chem Soc 123:6461
273. Kanaoka S, Sawamoto M, Higashimura T (1991) Macromolecules 24:2309
274. Sawamoto M, Kanaoka S, Higashimura T (1999) Star-branched functional polymers by living cationic polymerization. In: Sasabe H (ed) Hyper-Structured Molecules I: Chemistry, Physics and Applications. Gordon and Breach Science Publisher, Amsterdam, pp 43–61
275. Asthana S, Kennedy JP (1999) J Polym Sci Part A Polym Chem 37:2235
276. Shibata T, Kanaoka S, Aoshima S (2006) J Am Chem Soc 128:7497

277. Kanaoka S, Yagi N, Fukuyama Y, Aoshima S, Tsunoyama H, Tsukuda T, Sakurai H (2007) J Am Chem Soc 129:2060
278. Okabe S, Seno K, Kanaoka S, Aoshima S, Shibayama M (2006) Macromolecules 39:1592
279. Seno K, Kanaoka S, Aoshima S (2006) Polym Prepr Jpn 55:358

Editor: S. Kobayashi

Author Index Volumes 201–210

Author Index Volumes 1–100 see Volume 100
Author Index Volumes 101–200 see Volume 200

Subject Index

The manufacturer's authorised representative in the EU is Springer Nature Customer Service Centre GmbH, Europaplatz 3, 69115 Heidelberg, Germany. If you have any concerns regarding our products, please contact ProductSafety@springernature.com

Printed and bound by CPI Group (UK) Ltd, Croydon, CR0 4YY
02/10/2026
02227132-0002